To Jeanne —
 This is what
was like in Grandma Hill's day.
  Karla

Cimarron Chronicles

Saga of the Open Range

By

Carrie W. Schmoker Anshutz

and

M. W. (Doc) Anshutz

Copyright ©2003 by Ohnick Enterprises, Meade, Kansas 67864
(800)794-2356
www.backroomprinting.com
Manufactured in the U.S.A.

ISBN 0-9746222-0-6

Forward

by LaDonna Meyers

While clearing out my mother-in-law, Imogene Schmoker Meyers' attic a few years ago, I came across many dusty boxes that stored a variety of her treasures. The item which intrigued me the most was a stack of yellowing papers. As I started reading them it was evident that it had been typed decades ago, the ink of the typing now fading. Many of the letters in the words, along with a few complete words, had simply vanished from the old pages. The story told of my husband, Larry Meyers' grandfather and great grandfather, Will and Chris Schmoker, and was written by Carrie Schmoker, Will's sister. Carrie Schmoker Anshutz was the grandmother of Mary Anshutz Finney.

I began retyping the story to preserve the wonderful history of our family and the history of the area in which we live. Mary Finney was a great help in filling in some of the missing words and helping me find photos that supported her grandparent's story.

I hope that you will enjoy reading the events recorded in this book. Carrie and Doc have tried to paint a picture with their words of the lifestyle of their time.

Editor's Note:

This book was written by Carrie and Doc Anshutz in the 1930's. The language they used is unfamiliar at times, and very quaint by today's standards, but very much a part of who they were and the place from which they came.

We have tried to keep editing to a minimum in an effort to preserve this unique quality because we feel the way the story is told is as important as the story itself.

Introduction

It is the 31st of December 1935, and at the stroke of the midnight hour, bells will ring out everywhere in a wild and joyous medley of sound. The young people all will join in some expression of joy or wish for happiness to be brought by the New Year.

The worshipers will break into songs of Praise and Joy. The wild throngs in the streets, the banqueters and revelers, in every place and walk in life will break into wilder revelry, a people gone mad for joy.

Joy for what? Joy that the old year with all its mistakes and hardships, its sorrows and disappointments, is past and hope that the new one will bring only pleasantness, peace, prosperity, some heart's desire fulfilled, a chance to right wrongs, to live better and more unselfish lives?

Perhaps, I do not know. I only know that it is a custom and also with many it is customary at this time to make good resolutions.

But my husband and I sitting in our modest ranch home in what was once "No Man's Land," the home that we have lived in for more than forty-two years, my husband and I are alone. It is very quiet here. We can only imagine the crowds, the noise, the singing and the ringing of the New Year's bells throughout the land.

The three children who blessed our home, two sons and a daughter, are married and in homes of their own. There are four very dear grandchildren. It is of these we are thinking this New Year's Eve.

What can we leave to them as a heritage? Not material riches. No. Better perhaps a written account, some record of the fast disappearing generation of pioneers of our day and the time and surroundings so different from that of the on-coming generation in this the land of their birth.

For I came with my parents to this country as a child and I grew up in the ranch surroundings and traditions. My husband came here a young boy and grew up with the country.

For many years we had lamented the fact that no writer had seen fit to leave some record of the early days of this particular scope of country. Following the slaughter of the buffalo and the placing of the Indian tribes under government supervision the plains country, that had been for so many years the home of both the Indians and the buffalo, then became for a time the cattlemen's paradise.

This scope of country stretched from Canada on the north to the Gulf of Mexico on the south. Its western border was the foothills of the Rocky Mountains and on the east it included all the western part of

Kansas and Nebraska, Indian Territory and Texas. "No Man's Land" was a strip of land that lay in the very heart of this cattle country, 165 miles long and about 34 mile wide. Its southern border was the Texas Panhandle, to the north of it was Kansas and Colorado, to the west lay New Mexico, its eastern border was the Indian Territory.

Well watered and finely grassed, it was considered one of the best cow countries in the West. Through some error, this strip for some years lay open without form of government and under no jurisdiction of any court. There were no laws, no taxes. Everything was free and open.

The cattlemen who first settled here were bound by no laws, but in the main they were a high class of men bound by a common code of honor and dissension was almost unknown in those days.

The cowboys who rode these ranges were a carefree, rollicking lot of young fellows. The hardships and discomforts of the life were passed over lightly by them, a rough and ready lot with the kindest hearts imaginable. They were ever ready to lend a hand to one less fortunate.

It had always seemed to us such a pity that so little written record was made of those interesting days. Now in the evening of life, as we sit in our quiet home, in what was once old "No Man's Land," we agreed that this should be our "good resolution" for the New Year. We will begin at once, now, to write the story of our West. A story not with the view of making it our life story, but rather as being typical, in a manner, of the lives of many with whom we were associated in the long past days. A story for their children and grandchildren and our own, we leave some record of the life we lived in those days.

"You write your story first," my husband said. "It will be the story of the first settlers and of the sheep ranch and when it is finished you shall write mine. It will be the story of the first cattle ranchers, the cowboys who rode the ranges joyous and carefree, and the cattle business that was such a thriving and important one for many a year."

Carrie Schmoker Anshutz

Carrie Schmoker Anshutz

Part One

My Pioneer Story

by Carrie Schmoker Anshutz

Chapter I

Since this is to be the story of pioneer life, I think it is but fitting to include a short sketch of the lives of my parents, as they were twice pioneers.

My father, Christian Schmoker, was born December 31, 1836, in the Swiss Alps in the valley of Gruendlewald. Here his ancestors had lived for generations tilling the soil of small patches of mountainside farmland to raise vegetables for the family's use and provender for the cows also. For every family in the village kept a few cows and goats and the milk from these was made into the famous Swiss cheese.

In summer the cows and goats were taken far up on the mountainsides and grazed there until the snows of early autumn compelled them to return to the valley. Each herd was accompanied by a boy of about fourteen years as herder and an old man as cheese maker. Together these two lived in log huts, milked the cows and goats and made and stored the cheese. The cheese was never considered fit to eat until it was at least six months old.

Occasionally some member of the family would come up to find how these two were faring and bring them some supplies but mainly they lived on milk and cheese.

Off on other mountain sides were similar groups with their herders and those boys in sight of each other but separated by yawning chasms and deep gorges. Too far apart to permit anything in the way of communications and yet eager for the sound of human voices, they developed a way of communicating. Standing high on rocky crags, opposite to each other, these boys yodeled. In the clear mountain air the sound carried for miles echoing and re-echoing against the rocky crags.

Father told us that some became so expert at this yodeling as to be almost marvelous.

Cimarron Chronicles

The family at home worked busily. Men and women alike worked in the fields, planting, cultivating and harvesting everything possible and storing for the long snowy winter. Far up on the mountainsides they climbed and cut grass for hay for winter use, carrying it down to the log barns on their backs.

With the coming of winter their mode of life changed. The cows were housed in the good warm log barns and carefully fed and tended. The children were sent to school. It was compulsory that all should attend school until fourteen years of age and it was one of these odd customs of that time that each pupil must carry a stick of wood from the home supply each morning to the school house. In that way the fuel supply was kept up.

The elders at home were occupied in spinning, weaving, knitting and all the homely tasks of life for in those days, one hundred years ago, almost everything was made by hand.

Hides were tanned and from these the family footwear was fashioned by the shoemaker who went from house to house during the winter staying at each home until the entire family was shod.

But for the young folks at least it was not all work now for the long winter evenings were passed in merrymaking and jollity unawed by the majesty of the towering "Matter horn" and the "Jungfrau."

Father said that they as children could never understand the enthusiasm and delight of the crowds of summer tourists as they viewed these mountain monarchs. In fact the Swiss youngsters regarded these tourists as a bit foolish in the head to come so far to see and rave about the mountains that to them seemed so commonplace.

Only in later years did they appreciate and long for another sight of the grandeur that surrounded their childhood home. When my father was about fifteen years of age (October 6, 1851), the family decided to immigrate to America. And so the village home, the small patches of farmland were sold along with the cherished cows and the faithful goats. The personal belongings packed and the family, father and mother with their five boys and two girls journeyed to Berne, from thence to Havre, France where they took passage on a sailing vessel for America.

They were six weeks on the waters and when within sight of land a terrible storm came on and the Captain of the vessel put back to sea. The weary passengers grumbled and protested to no avail. The ship was driven so far out by the gale that it took two weeks to get back. However, the sight of several vessels that had been wrecked in trying to make harbor during the storm completely silenced the ones who had complained so bitterly.

Part 1 • Chapter 1

Our family landed at New Orleans and then came up the Mississippi by boat stopping a short while at St. Louis. They then went on up the river to Davenport, Iowa, where search for a new home began. Soon after arriving at Davenport the father was stricken with pneumonia and passed away leaving his family desolate among strangers in a strange land.

Land was purchased in Muscatine County, Iowa. In the hills and wooded bluffs a few miles back from the mighty Mississippi, the Schmoker family made a home amongst other pioneers who were clearing out timber and making tillable fields.

Here the children grew to maturity, married and acquired homes of their own in the community along Pine Creek, a tributary of the nearby river.

My mother, Anna Wunderlich, was born in Bavaria May 15, 1843, in a great old castle on the banks of the Rhine. The owner of the castle was often away for long periods of time. Grandmother was housekeeper and Grandfather was the caretaker of the place.

When mother was four years old, in 1847, the family came to America and they too had a long hard voyage in a sailing vessel. The tragedy of their voyage was the death of the baby boy of the family, a child two years of age. His burial at sea, when Grandmother frantic with grief, had to be forcibly restrained from leaping overboard when the little body was lowered into its watery grave.

The family lived a year in Muscatine and then moved to a farm on Pine Creek that was then mostly woodland and mother said lonely beyond words. This place now borders on one of eastern Iowa's most beautiful and popular State Parks. "Wild Cats Den" State Park is only a half mile from my mother's childhood home.

However, new settlers kept coming in. Land was cleared of timber and before many years had passed, a typical country community had developed.

Photos taken in 1995 of the Jacob Schmoker Chalet in the Swiss Alps in the valley of Gruendlewald. The house is 200 years old.

Chapter II

Soon after my Father, Christian Schmoker, and Mother, Anna Wunderlich, were married on September 20, 1860, they settled on the farm that was my childhood home. A little farm house perched on a hill. It still stands just as it did when we children romped and played and quarreled in and about it.

It was there we hung up our stockings on Christmas Eve with thrilled thoughts of Santa's coming. It was from there we trudged through snow and slush all during winter to the little country schoolhouse a mile and a half away. There to be taught our three "R's," "Readin and Ritin and Rithmetic." In those days they were all too literally "taught to the tune of a Hickory Stick," for as I remember them most of the teachers were young men with very short tempers who seemed to think that even the smallest infraction of rules must be followed by a good "licking." But these ill-tempered men teachers were later replaced by a very lovely and understanding young lady who taught all the later terms of my time in school. All the boys fell in love with her and the girls all adored her so there was no trouble at all and our school days were happy ones.

There was always a summer term and a winter term of school. The summer term being patronized only by the small children of the community for the older ones had work to do at home.

I disliked this summer term for I had to walk that mile and a half all alone most of the time. I was terribly frightened sometimes by seeing tramps and always imagined wild beasts were in the woods that bordered the roadside most of the way.

But the winter term was fun, for then all the older children went too, a whole bunch to walk together to and from school. The woods were naked then and did not hide wild animals as I imagined they did in the dense shade of summer foliage.

In winter our schoolhouse was filled to bursting with fifty to sixty pupils. There were all sorts of exciting games at noon and recess times. There were snow fights, Black-man, Dare-base, Whip crack,

Ante-over and many others. And we had all sorts of school activities for the winter evenings. Spelling matches, writing school, singing school and most exciting of all, school exhibitions or entertainments when every pupil had to perform some part for the benefit of proud parents and admiring friends.

Our community life was centered around the school and the churches, of the latter there were two within less than a mile, Methodist and Congregational. Mother's people were members of the Methodist Church, Father's the Congregational and it was in this faith that we were brought up. The people of the community were mainly Germans and our services were in German. Sunday school and preaching each Sabbath morning and it was in this Sunday school that I learned to read the German language.

But there were a number of English speaking families scattered about the community so for their benefit English Sunday school services were conducted each Sunday afternoon in our little white Congregational Church. My Uncle John Schmoker was the Superintendent of this and Father was his assistant. The attendance was splendid for the church was generally well filled with grown young people many of whom walked three or four miles to attend. We children seldom missed attendance at both forenoon and afternoon services, but I liked the afternoon ones best for many of the young folks were good singers and the song service was my delight.

Always at Christmas time we had a tree from nearby Pine Creek, a straight and beautiful pine, whose tips touched the ceiling and its beauty when the candles were lighted was the marvel and delight of our young lives. And each year on the Fourth of July the Sunday school had a picnic in some nearby grove. Barrels of iced lemonade, ice cream, and the usual picnic dinner and heaps of fun with swings, firecrackers and other devices made the day a delight to all the youngsters.

Those were our simple pleasures. We almost never went to town nor indeed outside of our own little community, shut in by its hills and woods. And suddenly, or so it seemed to me, my Father and Uncle John decided that it would be best to move "Out West" where their boys would have a chance to grow up with the country and acquire lands and homes of their own, where land was cheap and easy to get. So early in the year 1879, Uncle John Schmoker came west to select homesteads for both families.

He went first to Clarenden, Texas, but for some reason did not like that country and so came north to Kansas.

Part 1 • Chapter 2

As there were no north and south railroads in those days he came up from Texas with one of Charles Goodnight's cattle herds being driven to Dodge City. And on his return trip home, we children were thrilled by the stories he told us of his trip with the cowboys who all carried six shooters, of the great herds of long horned cattle, of cooking over camp fires and sleeping on the ground.

The homesteads he located were some fifty miles south and west of Dodge City in the Crooked Creek Valley. The United States Land Offices for this district were then in Larned, Kansas. The filing fees were $14.00 and a five years residence on the land entitled the homesteader to take 160 acres of land filed on. This seemed a wonderful opportunity to acquire, almost free of charge, a quarter section of land that lay almost level as a floor and of fine rich soil.

What visions those early homesteaders had of wonderful crops and fine homes that would be theirs in just a few years. How different were the realities of the new life they entered upon with such bright hopes and unfaltering confidence. But as yet all of this was in the veiled future, so in great excitement and high anticipation, at least on the part of the children of the families, we began preparations for moving. We could scarcely understand why our mothers did not seem to share our joy and enthusiasm.

Soon the farms were sold, ours to a German and his wife, neither of whom could speak a word of English. They hailed from somewhere near Davenport, and when they came to complete the land transaction they drove down to our place in their farm wagon, bringing the cash, $2,500. And as protection against robbers they carried with them two stout hickory clubs. The son, an only child of these people, still lives on the farm, an eccentric old bachelor.

A public sale was held and our livestock and farm implements disposed of. Huge boxes were packed with clothing, bedding and all necessary household utensils. A freight car was chartered and loaded. It contained, besides our personal and household goods, new wagons, plows, cultivators, harrows and other farm implements. After these were stowed in the car, shelled corn was poured over all, filling every nook and crevice, making all secure and firm. In one end was a compartment that held our good mules and two large crates of chickens. Potatoes, dried apples, sorghum and other farm products were also included and plenty of garden seeds.

Father and cousin Ed Schmoker came out with the freight car to look after the stock while Uncle John undertook the job of herding the rest of us through on the passenger train. There were nine children in Uncle John's family, four boys and five girls and in our family six children, three girls and three boys. In both families the ages ranged from seventeen to two years, counting out the boy who came in the freight car with father, that left two mothers, fourteen children and one father. So it was no wonder that one place where we had to change cars that we heard an astonished onlooker exclaim, "There goes another Mormon family."

We, ourselves, realized how ridiculous we must appear with our numerous bags and bundles for there were no dining cars and we had to carry with us food to last us the trip. Each of the older ones, who were not loaded down with luggage, had three or four of the younger children to look after, that none be left behind or stray away. Almost none of us children had ever been on a train in all our lives so it was an experience with a capital E. It was thrilling!

To be sure we felt a little sadness at parting from all our young friends, schoolmates, cousins, aunts and uncles, besides two aged and well loved grandmothers, whom we children never saw again. And the memory persisted of that last Sunday afternoon when the little white Church was crowded to capacity with all the young friends from far and near who had gathered to bid us a tearful and affectionate goodbye. We left Muscatine, Iowa about four P.M. of a day late in April 1879.

Those of us who were fortunate enough to get next to a window watched the Iowa landscape glide by until darkness settled down and we slept fitfully by snatches sitting upright in the seats.

Next day through Missouri, the country still had a familiar appearance of farm homes and small towns, but by evening of that day someone said, "Now we are in Kansas." And gradually the old order of things passed away for even in eastern Kansas, at that early day, farm homes were far apart. There seemed to be vast spaces of just grassy land with little wooded streams here and there, strange and lonely looking, but lovely in the fresh green of early spring. And I wondered why Uncle John had not found us a home here instead of going farther on. But I reflected perhaps, after all, the place we were going to would be better than this, for such is the optimism of youth. I was then ten and one-half years old.

And so, with my nose pressed flat against the windowpane, I watched my first Kansas Sunset. At home in Iowa the sun had always just dropped casually out of sight over the big hill, a half-mile west of our house. But this was my first experience of watching the sun set on a far horizon, the afterglow of crimson and gold and then the lengthening shadow, purplish, alluring, growing ever darker and more mysterious. I watched until I could no longer distinguish objects and then another night slipped away, much faster than the preceding one had, and the next morning we arrived at Dodge City, "The Cowboy Capitol of the West."

William Schmoker

Ella Schmoker

Chapter III

A short distance from the Santa Fe depot and south of the railroad was a modest boarding house of unpainted pine, kept by a middle-aged widow, a Mrs. Lewis. To this place we were taken, glad of the chance to stretch weary limbs and get some exercise, warm meals and beds to sleep in, although most of us youngsters were bedded down on pallets on the floor in the large downstairs sitting room that adjoined the bedroom occupied by our mothers.

As the freight car had not yet arrived we knew that we must stay in these quarters for several days. Time, being heavily on our young restless heads and minds, for a strict parental "edict" had been issued that not one of us should stir out of sight or hearing of our watchful mothers and this was strictly enforced.

Mrs. Lewis' help consisted of two strapping young fellows, not at all tidy as to looks, who answered to the names of "Kid" and "Rube." I think that after their days work was done these two must have spent most of the night about town for every morning Mrs. Lewis stood at the foot of the stairs and called and called their names. First one and then the other until finally sleepy and dissipated looking, they came stumbling down the stairs. Rube was the cook and Kid waited on the tables, long tables roughly made and oil cloth covered and for chairs, long benches.

Three times a day we were served buffalo meat, bread and coffee and not much else. The buffalo steaks were very dark and had a sweetish taste just as if they had been generously sprinkled with sugar instead of salt and pepper and unmistakably there was a sprinkling of sand for it gritted in our teeth at almost every bite. I went on a hunger strike but that is all the good it did me. Later I became accustomed to buffalo meat and when Mother cooked it, minus the sand, it was very good.

We made the acquaintance of one little girl during our stay with Mrs. Lewis. She lived nearby and seeing swarms of children at the boarding house came to call on us. This fortunate creature was not restricted in her goings and comings, but went her joyous way where and when she pleased. Our mothers did not wholly approve of her, but could not well banish her from our midst and she managed to impart to us considerable information for she seemed to know everyone in that part of town.

Close by in a little draw was a tiny dugout, the first I had ever seen. A crippled man lived in it all alone. He was a young man but rough and unkempt looking and he walked on his knees, which were leather protected. We were told that he had been caught out in the plains in a terrible blizzard and his feet, so badly frozen, that amputation was necessary. I was afraid of him but our young informant assured me that he was a nice fellow and a good friend of hers.

Later in the day we watched with interest a number of horsemen ride up to the large livery barn a short distance west of the boarding house. These, the little girl told us, were cowboys.

After what seemed to us a long wait, the freight car got in. Then all was bustle and excitement for it had to be unloaded, the wagons set up and everything made ready for the trip out to our homesteads. The car was left on a rising some distance back from the depot and now, joy of joys, we were permitted to go out there and watch the men and boys at their work and our spirits rose accordingly.

It was here that we made the acquaintance of the colony from Zanesville, Ohio, who were going to settle within a few miles of the homestead we were going to.

They had come from Zanesville in an emigrant car. All of them were city people without the least experience of country life, pale, anemic looking young men who had been clerks and office workers, and their equally frail looking wives. All had been lured to this new wild country by the magic offer of 160 acres of land, almost free of charge.

Some of these who had never owned an acre of land were sure this meant riches. The men were enthusiastic in their outlook on the possibilities in store for them. But the women, tired from their journey, sitting there in the open railroad yards on their trunks and boxes, waiting for the husbands to find some means of transportation to their waiting homesteads, looked about them and voiced their doubts as to the wisdom of their move. To them the new country looked very lonely and bare and they were not enjoying the hot sun and the wind that blew steadily from the southwest. They spoke of the lack of shade trees and the stark ugliness of the frontier town, for the Dodge City of that day, was a rough looking place.

Part 1 • Chapter 3

The buildings, mostly of unpainted pine, were roughly constructed and a few board sidewalks along the Front Street, where all the business houses were. The streets and roadways were dusty, sandy trails; no well kept yards with trees and shrubs and flowers, nothing attractive anywhere that I could see. So we were glad, when on the fourth day after we arrived, that things were in readiness for leaving there.

Two freighters with lead and trail wagons had been hired, their wagons and our own piled high above the sideboards with our goods. Each wagon bearing its quota of human freight disposed of to safest advantage topping it all, we started south and west on the wide Jones Plumber Trail. Crossing the Arkansas River on the wooden toll bridge that had a wooden bar across it at the south end, where the toll house stood, each wagon was charged a toll of 50 cents for a single team, a six horse team was one dollar, horsemen were charged 25 cents.

We got a late start and the going was slow so that night we camped about twelve miles out from Dodge City on Mulberry Creek. Our campsite was on a sloping hillside a short distance back from the creek and I was horrified to see that the place was thickly strewn with the carcasses of sheep that had perished there in a snowstorm of the previous winter. "We are not going to stop at this nasty place to eat and sleep," I wailed. And I was promptly told to hold my tongue and that my elders would make plans and decisions for me. In fact, although I did not know it, this was the only place on the trail where water was available until some forty miles farther on for it was a dry season.

The freighters took charge of the camp arrangements. They built a fire of the dry "buffalo chips" or dung, and over it boiled the coffee and fried bacon and dried antelope ham that with bread made our evening meal. One of the freighters, on going to the creek for water, asked some of us smaller children to go along. We did so and I was further nauseated by seeing more sheep carcasses in the creek bed, but I said nothing, merely deciding in my own mind that I would refrain from drinking and this I did, wondering how the others could possibly stomach it.

Supper over, a wagon sheet was stretched along the side of a wagon and staked to the ground on the further side. Thus making a sort of shelter where bedding was spread on the ground in a double row like sardines in a can, and slept? No, passed the night away, at least some of us did only that. Unfortunately for me, I was placed in the lower row and we lay down fully clothed even as to the shoes. The cousin who was place just above me slept quite soundly, but it was a restless sleep and the heels of her copper-toed shoes beat an almost constant tat-tat upon my defenseless head.

With the morning and we were again on our way. And now the trail led over the great flat that stretched away mile after mile, as far as the eye could reach. Not a tree, not a habitation, nothing in sight, but just grass and sky and the wide hard beaten trail. At times our way led past prairie dog towns and the little scurrying beasts standing up with front paws folded like arms across their breasts, barking saucily, made a welcome break in the monotony of our slow tiresome journey.

Toward evening of that day we came to a little 8 X 10 shack of rough pine, newly built, and a young man came out to greet us. He said his name was Harvey Penrose and his parents and sisters were even then on their way out to this place from their former home in Iowa. This was indeed cheering for it meant that we were to have neighbors.

Uncle John said we were now only a couple of miles from our homesteads and we left the trail and struck off west across the unbroken prairie to the place that was to be our home. The freighters camped with us that night. The next morning after the wagons were unloaded they bade us goodbye, wished us luck, and drove away leaving us alone in that trackless waste. I distinctly remember the feeling of almost panic as we all stood and watched them drive away. They were friendly young fellows, well acquainted with the country, and we were so new and so "green." I wished they might stay with us longer until a little of that awful feeling of desolation had time to wear away.

We had brought with us two brown canvas tents of good size and they were now set up and our belongings stowed within. Beds were set up and what luxury it seemed to sleep in a real bed once more. A cook stove was set up in a sort of lean-to arrangement on the north side of one of the tents. But our mothers had their trials with that stove for the wind blew down the short length of stovepipe it carried and caused it to smoke. And they had to become used to the fuel, dried buffalo chips that burned up in such a short time it kept one busy stuffing the stove and taking out the quantities of ashes that resulted. Quite different from the nice clean wood fires they had always had in Iowa.

The freighters had carried with them a supply of water in kegs, now we must find a water supply of our own. My Uncle John knew there was a spring somewhere nearby but did not know its exact location, so now men and boys started out on a search for this spring. They were gone such a long time that some of us small girls decided we might as well help in the search and started off in the direction of the creek, but had not gone far when we met one of my cousins returning. He was shocked and angry to find us wandering off from camp and gave us a real lecture on the danger of getting lost on the prairie, of being bitten by rattlesnakes. And besides, he said, there were wild and terrible cowboys who carried guns and six-shooters and we might even be found and carried away by Indians. Oh, it was a lecture we did not soon forget and although it was all exaggerated, perhaps, it was just as well for thereafter we took more heed, at least for awhile.

The spring was found about a mile west of us, an abundance of clear sweet water. We called it the Lockhart Spring for a family by that name had homesteaded the place it was on and soon came to live there in a dugout in the bank just above the spring.

All that long, hot dry summer we made daily trips to the spring for our supply of water that had to be hauled in barrels. Sleds were made to haul the barrels of water on, and the runners soon became slick and slipped easily over the buffalo grass sod. The lack of cool fresh water was one of the hardships of those days for the water in the barrels was always warm and our trips to the spring were the only times we ever had good cool water to drink. But that was not the only pleasure of going to the spring, for it was an oasis of green and a lovely spot, surrounded by sun burnt prairie. Tall green grasses grew along the little stream that flowed away from the spring and lovely deep red flowers grew there in rank abundance.

Here the moist earth, the flowers and foliage gave forth a rich odor that I remember to this day. Meadowlarks sang joyously all about the place so we children delighted in going there. The water sleds usually carried a number of passengers going, but coming back we had to walk for the water 'sloshed' out badly sometimes when the sled struck a rough spot and the barrels were load enough, but we were used to walking and liked it.

Cimarron Chronicles

Chapter IV

The first Sunday on our homestead I remember very well. After dinner the mules were hitched to the wagons and we all drove down to see the salt well that was then very new and mysterious. The freighters had told about it. It was five miles south of our places, just off the Jones and Plummer Trail and had fallen about two months before.

A freighter going south from Dodge City with his loaded wagons had camped almost on the spot and coming back a week or so later found, right across the track that his loaded wagons had left, a yawning chasm, a circular hole thirty feet deep and two or three hundred feet across and at the bottom of this a pool of heavy salt water. The ground all around for yards back was seamed with deep cracks so that we approached cautiously and peered down into the blue depths, fearing that at any moment more of the ground would cave in taking us all along with it. But our fears were entirely unfounded for the salt well continued to be a point of great interest for a number of years.

Soon a well-worn path led down to the waters edge and the place became a favorite swimming hole for all the boys and men. In the early 1880's an enterprising man saw the possibilities for making salt. Many long, wide wooden troughs were made and into these, by means of a pump operated by a steam engine, the water was drawn and allowed to evaporate leaving a residue of pure salt. However, there was little demand for it locally, as stock salt was not used to any extent in those days and the nearest shipping point was Dodge City, so the project was abandoned.

I believe it was later established that an underground stream of water, probably the artesian, flowed over all that part of the Crooked Creek Valley. Flowing water over a bed of salt washed out a cavity here and the heavy trail wagons loosened the top crust causing it to fall in, for the old Trail from Fort Dodge, Kansas, to Fort Bascom, New Mexico, (near Tucumcari) lay directly over the spot.

When we were there the trail marks were new and distinct on both sides of the chasm and we shuddered to think of what might have happened had a freight outfit been over the spot when it collapsed.

The water was very deep at first and so intensely salty that it had a blue-green look. As the years went on and on and no water was pumped out, the lower opening must have filled with silt washed in from the surface so that gradually the pool and the bottom of the sink diminished in size and now is quite insignificant looking.

As soon as possible after getting our tents settled our teams were put to work hauling out lumber from Dodge City and houses were built a short distance apart. Story and half frame buildings about 16 x 18 with lean-to kitchens added later. I think it was about two months before the houses were finished so that we could move in. It certainly seemed good to have a roof over our heads once more and stoves that could draw without smoking, for the cooking had been a task while in the tents. There were such a lot of us and we had developed such enormous appetites in this clean sweet upland air.

One evening when the wind was blowing a perfect gale, there was no bread baked and it was impossible to get the oven heated for biscuits. We had just cornmeal mush for supper and nothing else, no milk to go with it. It seemed to stick in my throat and I have never cared much for mush since.

Although there were many threatening looking clouds that spring and summer there was no rain, not even one good shower. Sometimes a few big drops came spattering down and the sky would be clear again. The weather was a puzzle to us. In Iowa when such black clouds came up it always rained. A garden spot was plowed, the sod chopped and worked fine, and seeds were planted, but not one of these ever so much as sprouted. It was a great disappointment especially to our mothers when this good black, rich looking soil would not produce the vegetables we had been accustomed to having in such abundance. With no milk, no vegetables, no fresh fruit, cooking was indeed a problem.

It was fortunate that we had brought chickens for the hens furnished us with plenty of eggs. But bacon, beans, bread, molasses and dried apples were the staples in those days, even potatoes were a luxury then. In winter when we could have fresh meat, it was not so bad. That first winter, buffalo hunters came from the south and sold buffalo meat to the new settlers and sometimes the boys were able to shoot antelope. Stewed antelope was very good as were also the antelope steaks.

Part 1 • Chapter 4

Several times the boys found young antelopes asleep on the prairie. They ran them down and brought them home. These lovely creatures we children tried to raise as pets but they invariably died after a week or two. The young deer were hardier and one of these my cousin raised and was his pet for many years. He was a beautiful creature and would eat anything a dog would eat; bread, pancakes, meat and many other things and in spite of this unnatural fare lived and thrived.

Many large herds of cattle came by our place that summer, being driven up from Texas to Dodge City. The cowboys with these herds usually rode up to chat a while, invariable remarking that it had been months since they had seen women and children. Generally, too, they tried to engage the younger of us children in conversation but we were too shy. We had heard too many stories of the wild ways of the cowboys to make friends or feel at ease in their presence.

The herds were typical old time Texas long horns of every variety of color and spot combinations and nearly every herd had with them a sprinkling of buffalo yearlings, mostly calves, that had fallen in with the herd on the way up.

New settlers came in surprising numbers that summer. The Ohio colony was about five or six miles north of us on Crooked Creek. Four of the families whose homesteads cornered made houses half dugout, half frame in a close group. And a store and post office was established. The place was called Pearlette and it was there that we got our mail for some time.

Most of these colonists, however, did not stay long for they were city people and with very little cash to go on they were compelled to find subsistence elsewhere. A few hung on for several years, but of them only a couple of families remained permanently.

Nearly all of the first settlers located along the Crooked Creek Valley or closely adjacent to it. A couple of miles north of us were the Eliason family, Norwegians, and soon after several other of their countrymen also came. The Reep, Peterson and Erikson families and a little to the northwest was another group, Milligan, French, Worth, Norman and Sourbier families.

A post office was established there later on and named Belle Meade, Cap Milligan was the Post Master. Farther along north in the Creek Valley was another group, the Conrad, Coffee, Haywood and McCanly families.

Our nearest neighbors were the Colgan boys, bachelors, the Lockhart family and two Penrose families. In a surprisingly short time we had made the acquaintance of them all for very soon a Sunday School was started and held each Sunday afternoon at the Sourbier home. The attendance was good, all being glad of the opportunity to meet and become acquainted; all drawn together by a common cause that of creating new homes in the vast reaches of prairie by which we were surrounded.

The Penrose families were most devout Quakers and each Sabbath day held their meetings at the Isaac Penrose home, a frame building much the same size as our homes. Soon they invited us to come and worship with them, so as it was much nearer home, we did so. It was arranged that we should have our Sunday School first, the lesson as arranged by David C. Cook Publishing Company and then after that the Quaker meeting which consisted of sitting perfectly still with arms folded and eyes cast down in deep meditation, presumably of things spiritual, and for the good of one's soul.

On rare occasions one of the other of the Penrose men "moved by the Spirit" would suddenly arise and make a short talk or a prayer. But most often we just sat, so still one might have heard a pin drop, for thirty minutes at least, sometimes longer. It just depended on how deep were the meditations of Isaac, who was the leader. Then just when it seemed to us children that we could stand it no longer, Isaac Penrose would stir, unfold his arms, raise his head and with a smile of friendly fellowship, shake hands with the person nearest him. The older people shook hands all around and the meeting was over. I am sure those Quaker meetings were fine, for us children we always felt so good when they were over.

Even though it was quite an ordeal for restless youngsters to sit quiet during theses meetings, we did appreciate and enjoy those two families as neighbors. They were educated and refined and the young folks were full of fun.

Isaac's family consisted of one grown son and four daughters, all grown except the youngest who was my age. Rachel was the eldest, a blonde, very quiet and reserved. Sue was a sparkling handsome brunette, Hannah was called the "Tomboy" of the family and a fair little May. There were four children in the Will Penrose family, the oldest a girl of eleven, the others ranging down to a baby.

After a couple of years of life in Kansas these two families decided to return to Iowa, partly on account of the failing health of Rachael who died sometime after their return to their former home. The vivacious and beautiful Sue, it was said, married the son of an Iowa governor; of the others we had no further knowledge. We were sincerely sorry to lose these fine people from our community life.

Part 1 • Chapter 4

In the autumn of that first year two more families came from our own Iowa community. The Peter Blair's with their three children, and my Aunt Elizabeth Schmoker a widow (of Peter) with two children, Abe aged twelve and Anna, aged ten.

They filed on homesteads adjoining Uncle John's and ours and made a welcome addition to our midst. The Blair's with the help of our men folks put up a substantial sod house that served them as a comfortable dwelling for many years. Aunt Elizabeth's was a frame house like ours.

While the settlement of the Crooked Creek Valley was going on cattle ranches were being established also, but these were far apart.

George Emerson had his headquarters ranch about twelve miles north of us on the creek, which was then known as the Emerson Grove. A fine grove of natural timber, the only one of any size in all the country around about, although on some of the little creeks and sheltered canyons there was then quite a sprinkling of timber.

Mr. Emerson had a summer camp on the creek valley a few miles to the south of us and his cowboys were holding about four hundred head of cows for some months. These were the first cowboys we had become acquainted with. George Decow was a reckless young daredevil, always pulling off some stunt. "Spike" Rexford, Billy Defreese and several others who became frequent visitors in our homes and were so friendly and full of fun that we children soon forgot to be afraid of them because they were cowboys

George Decow soon married and settled on a place in the settlement, and Billy Defreese also settled on a place of his own. But poor Billy, so genial and fun loving, with the most infectious laugh I have ever heard, met an untimely death a few years later in the first terrible blizzard that the new settlers experienced, and in which a number lost their lives but of that I will write later.

The first summer in our new home seemed to us children more like a prolonged holiday. There was little work we could do so most of our time was spent in play. But there was one thing that our mothers insisted on and that was that an hour or two each day must be spent in learning to sew. With an enthusiasm that later lagged we all started patchwork quilts. Mine, when finished, was far from being a thing of beauty, for it was crooked and puckered in places, but it served a purpose and in time we all learned to sew well.

After the buildings were put up our fathers and the boys had leisure time also. They made several trips in the surrounding country just to get the lay of the land. On one of these trips my older sister Elizabeth and Cousin Mary went with the men on a day-long trip over in the rough Sand Creek country. In order to measure the distance traversed they tied a white handkerchief around one of the wagon wheel spokes and then took turns counting the wheel revolutions.

At another time I was included in the party that went after wild plums, over in the rough hills and along a little creek, where we found them in abundance and gathered several bushels to take home with us. Very few of them were cooked for we ate them raw. That was the only fresh fruit we had that year and for many following years. Although we did sometimes find wild grapes, these were so sour and seedy that they were never looked on with much favor unless made into jam or jelly but sugar was much too scarce to allow much in that line.

It was late summer or early fall when father bought a couple of ponies. I think Uncle John bought some at the same time for all of us were eager to ride. One of ours was a low, heavy-set brown mare we named Nellie and the other was a horse we called Blue because that was his color. Both of them were good and gentle, but we only had them a short time when Blue was bitten by a rattlesnake and died. That was certainly a blow for us. We had planned so much on horse back riding and now only had Nellie left and there were such a lot of us.

But one never to be forgotten day, Nellie was saddled and had to be taken to the spring for her daily drink. I happened to be at the barn and begged to be allowed to ride to the spring. Brother Will, good naturedly assented, boosted me into the saddle, tucked my bare feet into the straps just above the stirrups, and placing the bridle reins in one hand he bade me hold onto the saddle horn with the other. As we started off he slapped Nellie's rump and she swung into a long steady lope that never once slackened or faltered all of the mile or more we had to go. I had never been on a horse but once or twice before so I knew nothing about guiding a horse. Afraid to let go of the saddle horn for fear of falling off, I could not pull her up with one hand for Nellie was hard mouthed. So we galloped on and on, in fact as we neared the spring Nellie went faster and faster for she was thirsty. Then on reaching the water she stopped so short I nearly went off over her head. Weeping bitterly, I got off and after she had her fill of water, led her home for both my legs were rubbed raw on those stirrup straps. It was a long time until I asked for another ride.

For the new settlers the only means of earning a few dollars was by gathering the buffalo bones that lay bleaching everywhere on the prairies and hauling them to Dodge City where a ton of them brought seven or eight dollars. Nearly all availed themselves of this opportunity, some made a business of it and for some time this was their only means of subsistence. It soon became evident to our fathers that they must turn to stock raising if they meant to stay with the country.

Cattle were very cheap at that time, good young cows and heifers could be bought for eight dollars per head. But they felt that they were not equipped, nor had they the experience to cope with the open range method of handling cattle. So they decided on sheep and accordingly each of them bought a flock of sheep and sheep ranches were established. Uncle John's ranch was on skull hill, east and south of the homestead, our ranch was lower down on Crooked Creek where we later moved and made our home.

The weather that winter of 1879-1880, was even more of a surprise to us than the summer had been. It was not winter at all, for the weather was so mild and sunny we went about most of the time without coats and outer wraps. Seldom was there a cloudy day. On Thanksgiving night we had a snowfall, so light it scarcely covered the ground, and all of it faded away by noon of the next day. I remember that quite distinctly for it was associated in my mind with the exciting event of a party for the young folks at the Penrose home that evening, a taffy pull, I think it was.

I was not considered old enough (12 years) to be included in the invitation and I watched the older ones drive off to the party with a feeling of disappointment not unmixed with envy. At holiday time there was a party at our house and Uncle John's. Billy Defreese persuaded our mothers to have it and offered to get a turkey. He did so, although where he found a turkey I am sure I do not know, but I remember that he helped dress it and rode to all the homes far and near and gave the invitations. Pies and cakes were baked and great preparations made for a real "back home" dinner.

All of the guest assembled at Uncle John's home because they had an organ. Music and singing and various games made up the evening's entertainment. While the guests were assembling and getting acquainted, our mothers and several of us girls of the in-between age were busy at our house getting the "feast" ready. I think it must have been just that to all who were assembled there for the evening.

When the long table was set and everything ready, Cousin Anna and I went to the other house to call them and they came trooping down to our place. When the meal was over, Mr. Schultz sat back and said, "Well, I never expected to eat a meal like this in this country. I have seen nothing like this since I left home." The party was a great success all around for the guests returned to Uncle John's home and all remained until after midnight.

A cold wind had blown up from the north by this time, so one party of young folks, who lived about twelve miles north of us were persuaded to stay all night at our house. It was still very cold and windy in the morning so they stayed on until afternoon when the sun came out bright and warm. So the ride home in the open wagon could be made with some comfort and by that time we felt quite well acquainted with these young folks whom we had never met before.

In the spring of 1880, there was some showers of rain, enough to start the grass off nice and green. Again we planted some gardens that never produced anything, the hot winds and sun soon burned everything up. The new farming implements still stood about unused, for that matter I might say right here, that they never were used but just rotted away.

We did manage though to have a few hardy flowers, zinnias and marigolds. Under the north window of the living room, protected from the hot south wind was a flowerbed where a few less hardy flowers that we watered and tended with great care were growing. There was one plant we called "Jacob's Ladder" for it grew higher and higher putting out flowers along its stem as it grew, brilliant orange red blossoms, almost every morning a new one had appeared during the night. One morning I dressed hastily and said to mother, "I am going to see how many new flowers have come out on Jacob's Ladder." But I did not go outside for the window was open and there were no screens.

I leaned out of the window and then drew back hastily for I had heard a sound, saw a slight movement and caught the gleam of a baleful eye. There coiled on the damp earth at the base of my cherished flower was a huge rattlesnake all ready to strike! The evening before one of my older brothers coming along the path from the barn to the house after dark heard a rattlesnake giving his warning rattle near the path. Reaching the house he warned us all to not go outside any more that night, saying it would be too dangerous to try to locate the snake that night but he would make a thorough search for it in the morning. As it happened, this was not necessary and the horrid thing was quickly disposed of to the great relief of us all.

Later when living on the sheep ranch we killed many rattlers for there was a prairie dog town nearby. One day I was going through "dog town" walking rapidly for I was on a hurried errand and I came upon one lying at the edge of a hole. Another step and I would have been on him. With a terrified leap I cleared snake and all, ran a little farther on and hurriedly looked back only to see him disappearing down a hole.

Part 1 • Chapter 4

I was glad of it for while we always killed them, if possible, and always carried a stout staff for that purpose. But on this occasion I felt weak and limp so that I doubt whether I could have accomplished the task. There is something about that warning buz-z-z of a rattlesnake, horses jump away from that sound with the same agility that I displayed then.

Cimarron Chronicles

Chapter V

It was in August of 1880, the 12th, that my brother Elmer was born. The only native Kansan of us all, our little Boy Blue, during his babyhood that was our pet name for him because of his big dark blue eyes. Two years later (Aug 24, 1882) Uncle John's family was increased by a baby boy, but Little Ben only lived about a year and had become inexpressibly dear to the family and then very quietly he slipped away.

In a little homemade casket, padded and lined with snowy white muslin, the tiny body was laid to rest. There in a corner of the homestead in a plot that already contained the grave of a stillborn child of the Blair's. Sweet, patient little Grandmother Blair later was placed beside the children and in the years that followed others of the settlers placed their dead in this plot.

When I last saw the place it was strongly fenced. But all of the families who had left loved ones there have moved away many years ago and I often wonder if time has entirely obliterated all traces of those graves. I could not find the spot, everything is so changed. I would be a stranger lost on what was once every inch familiar ground.

My Aunt Elizabeth and her two children stayed but a little over a year, her health was failing and she felt the children should be in good schools. Besides they found it very lonely and during that fall and winter I often stayed with them for days at a time to keep them company. We three children spent the long evenings playing dominoes and other games. Then we listened while Aunt Elizabeth told us stories of her childhood in Switzerland.

Daytimes we often heard many sharp reports sounding like pistols shots and then we always ran to a little rise of ground a quarter of a mile or so away from the house. Beyond that in a little depression that hid it from view of the house, the Jones and Plummer Trail lay and the sharp pistol like reports were made by the snapping of bull whips of the drivers of the ox trains. "Bull Whacker" these drivers were called and standing at a safe and respectful distance we watched the long "bull" trains as they slowly filed by on their way to far distant points.

In the early 1880's range cattle were turned loose to roam at will and these soon appeared in numbers along the Crooked Creek bottoms where the grass grew so rank and plentiful. When the first of these appeared my little sister Ella and two or three girl cousins had an experience they did not soon forget. They were walking to a neighbor's home a mile or two in distant and across the creek, which was quite dry as it had been ever since we had come to the country. When about half way between the two homes they passed near a bunch of grazing long horned cattle. At the sight of the small girls the animals all raised their heads, looked and then started toward them on a trot. In panic the girls fled for their lives, or so at least they thought. On and on they raced until they were ready to drop with fatigue and still the fierce long horns trotted after.

Frightened almost out of their wits, the children reached the shelter of a clump of tall sunflowers growing on the creek bank. Hidden by them they fell to the ground breathless and exhausted. The cattle standing off at a short distance watched and waited knowing full well that what they had followed lay hidden there. In vain the girls waited for the animals to leave. An hour passed, two hours and still the long horns kept their vigilant watch. The girls debated what to do, what could they do. "We will just have to lay here and starve," one of them said, "No one will ever find us here."

The oldest of them spoke up briskly and decisively. "Well, girls it looks as if we must die, but let us be bold and try once more to outwit those brutes. Now come, all together let us rush right out at them and yell with all our might." And yell and shout they did, rushing right into the faces of the astonished long horns that lost no time in turning tail and running like mad with the excited girls in hot pursuit. Never again were any of us afraid of the range cattle for we found that following us was mere curiosity on their part and to run at them and shout always put them to rout.

The Emerson Grove, twelve miles to the north of us, on Crooked Creek was an enticing spot. Here every 4th of July for a number of years the settlers gathered for the time honored celebration of that day. The only timbered tract of any size in the whole country side, its huge old cottonwood trees interspersed with willow, walnut, hackberry and china berry made a shady retreat from the burning July sun and everyone gathered there from far and near. Some came from as far away as Dodge City to the annual celebration. The most of them came in wagons, some on horseback, and a few in buggies or buckboards. It was a great social event in the community.

Part 1 • Chapter 5

If my memory serves me right, it was July 4th, 1880, that we had our first, really good, big general rainstorm. There was a large crowd at the grove and everyone enjoyed the day even though it was intensely hot, even in the dense shade of the grove. By that time we had become almost accustomed, though not quite reconciled to the heat, the drought and the hot winds.

The Blossom's had a sheep ranch at what is now known as Big Springs Ranch, southwest of Meade, Kansas. The family consisted of Mr. and Mrs. Blossom and a grown son of Mrs. Blossom by a former marriage. His name was Will Lee, a dark rather sullen looking chap, and adored by his mother. Mrs. Blossom was a tall, handsome woman, a Kentuckian, fearless and self-reliant, as all pioneer women should be. However, she found ranch life lonely and wanted a girl for company so mother let me go.

I was rather homesick at times, for I missed the other children so much. I had never been away from the home folks before, but there were compensations. For one thing, there was a great supply of reading matter there, stacks of novels and fashion magazines of that day. Always fond of reading, this was a great treat for me. Mrs. Blossom carefully picked out the books she considered fit for my young mind but I delved into all of them, or as many as I could.

Mr. Blossom usually spent the evenings reading aloud to us. The "Wandering Jew" was one of the books he read to us. I never forgot it, although at the time, I did not wholly understand it.

I spent many hours wandering along the Spring Creek. It was so pretty, a shallow stream flowing over a pebbly bottom and shaded by cottonwood trees in which many birds nested and sang joyously. Happy little songsters, many of which fell victims to the greed of "Rowdy," a beautiful Maltese cat that was Mrs. Blossom's special pet and pride. But it was the bane of my young life for regularly every night he came in through the window that was left open for him, carrying a live bird. Invariably, he brought his "catch" into my bedroom where he released the terrified thing, letting it fly and flutter about. Always catching it every now and then, teasing it as cats will, and at last I could hear him crunching and eating it.

Mrs. Blossom loved pets and sometimes found young mocking birds in the nest. Some of these she raised and taught to whistle tunes. By splitting their tongues, these birds could talk almost as well as a parrot. She had one that could talk quite plainly, whenever he saw a cat he would call, "Rowdy, Rowdy, look out, look out."

Being a Kentuckian, Mrs. Blossom loved good horses and knew how to handle and train them. She had a team of driving horses that were beauties and real steppers. Once every two weeks or so, they were hitched to the buckboard and we would drive up to the settlement for the mail. We always stopped at my home for a little visit and those were red letter days for me, for I was very lonely, at times, and wished with all my heart for the little sister and cousins who were my playmates at home.

After I had been at the ranch a short time Will Lee, the son, became tired of the lonely life of a sheepherder and decided he would try the more glamorous life of a cowboy. So much against his mother's wishes, he left home and for a time she did not know where he was and was greatly worried. However, by diligent inquiry he was finally located at a ranch about a days journey east of the home. Mrs. Blossom decided she would go to him and if possible persuade him to return home as winter was coming and there would be little chance for a green hand to be kept on at the cattle ranch during the winter months. So taking me with her one morning, we started for our unknown destination. She knew the general direction we must take but there were no roads to follow so we set off across country.

By noon we reached Uncle John's sheep ranch where we found my cousin Silas in charge of the herd and we stopped at his dark little dugout for something to eat. He set out for us what he had cooked, which was a pot of beans that had been prepared the day before and they were just slightly soured, so we had to count them as 'out' and that left biscuits, sorghum and coffee. After a rest and some further information as to our destination, we continued on our way to the east.

All afternoon we drove, mostly across rough country and without seeing a human being. Toward evening we came across several bunches of grazing cattle and this was an encouraging sign but as the sun sank low and still lower, we still had yet found no sign of the ranch we sought. Mrs. Blossom became worried and considered the possibility of having to spend the night in the open wilds. Then just as the sun was sinking from sight, we saw, coming in our direction, a solitary rider who proved to be a line rider from the outfit we were looking for. He was a friendly Texan and when Mrs. Blossom told him she was looking for her young son he told her Will was in camp and that he would show us the way in. Our meeting this cowboy was fortunate indeed, for we were still some distance from the camp that was but an incomplete dugout in the hillside near a little creek and it was after dark when we finally arrived.

Part 1 • Chapter 5

Great, indeed, must have been the astonishment of those men when a woman and a small girl suddenly appeared out of no where, in that place probably forty miles from any settlement and half that distance from any other ranch. However great their surprise, none of it was shown by word or look and we were received with the utmost courtesy. Everyone was very nice and expressed their pleasure in making the acquaintance of Will's mother, and the little girl received a good deal of attention, to her great embarrassment. At that time, I was a shy young creature.

Mrs. Blossom was a good talker and I am sure these men enjoyed the unexpected company as we all sat about the campfire chatting until bed time when they robbed themselves of enough blankets to make us a bed on the dirt floor of the small dugout. The men slept on the ground outside. Our bed was hard, oh very hard indeed, so we were glad when morning came. Outside the cook was getting breakfast over a campfire, the air of early autumn was crisp and sweet. As I stood looking about at my unfamiliar surroundings, I was suddenly quite dismayed to see the sun rising over the horizon's rim exactly in the direction I had fixed in my mind as being due south. And I could not get straightened out as to directions until we got back in more familiar territory.

I cannot recall the name of the cow outfit to which we paid this unlooked for visit, but the location was on some creek west of where Ashland, Kansas now stands. Breakfast was served us on a couple of tomato boxes for a table, the men making profuse apologies for the slimness of the fare as the wagon had not been in to Dodge City for supplies for a long tome and they were about out of everything except coffee, bacon and bread.

Our mission accomplished, for Will had promised to come home in a few days, we turned homeward once more accompanied by our guide of the night before. He now showed us a shorter and better way than the rough country we had wandered over the evening before and so our high stepping team carried us safely back home from what, to me at least, seemed a real adventurous journey. I think it was the following spring that the Blossoms sold out and left the country.

Cimarron Chronicles

CHAPTER VI

The only other sheep ranch in the country at that time besides the ones I have mentioned was the Sam Williams ranch on the Cimarron River. Mr. Williams was a middle-aged Scotchman, sandy haired, red whiskered, and a big, gruff and outspoken man, quite a character in his way. My brother George worked with him for a season and had many a story to tell us of "Old Man Williams" and his odd ways. He was a good camp cook and brother got lessons in that art which came in handy later on when the boys had to "bach" at our own sheep ranch. Our family continued to live on the homestead for some time longer while father went back and forth between the two places looking after them both.

I was curious to see where and how the boys were living so one day father took me with him on one of his trips down there. It seemed a long, long way across the high flat. As we jogged a long we saw just the wide expanse of luxuriant buffalo grass, no human beings in sight, no houses, no fences, nothing except a herd of antelope in the far distance. I was glad when after several hours in the monotonous travel we reached the hills that meant we were nearing Crooked Creek.

I shall never forget my first sight of the place that was later to be our home for many years. It was late in May and along the little spring fed branch, the meadow grass was so tall and luxuriant that the milk cows, Spot and Blue, were half hidden in it. Near the big spring was the dugout and in back of it, on a little flat, were the sheep corrals. The creek was about a quarter of a mile to the south and a few stately old cottonwoods grew near it. The creek, itself, was not at all like the Crooked Creek, which our homestead was on. For there its bed was a muddy bottom, dry except during much rainfall while here the creek flowed over a sandy bed and in places there were pebbly beaches that quite took my fancy. What a delightful place to wade and play if only the other children were here.

The ranch site had been selected because of the abundance of fine spring water, the shelter of the hills for the flock and last, but not least, for the very good dugout, which was all ready to be moved into. The

Lovell Cow outfit had made this dugout in 1877. They abandoned it in the following spring and some time later it had been used as a line camp by the Bullard Outfit who also left it and moved on. The dugout was all underground with three or four steps leading down into its dark depths. A fireplace was at the farther end and the floor and sides were of bare earth. A huge cottonwood ridge pole through the center of the roof on which smaller poles were laid on each side, these covered with rushes and then earth heaped over all made a strong, durable but ugly habitation.

The boys laughed at me for being afraid of snakes in the bunk beds that were built against the earthen wall. I tried to help get the meals but found the fireplace and the Dutch oven were entirely beyond my ability to handle. In fact the old dugout seemed quite awful to me and I pitied the boys for having to live in such a place, although they said they did not mind it at all.

It was not until later when we all moved to the ranch and into another house which father had made for us that we all learned the value of the old dugout. It was to its shelter we often fled during the awful windstorms that threatened to tear off the roof from over our heads. And in the first great blizzard which swept the country and caught us with very little firewood cut, father away and the boys busy with sheep, that my sister Elizabeth and I dragged poles, from the nearby woodpile, down into the old dugout. We got the sawhorses and the bucksaw and took turns sawing wood until we were quite exhausted but we kept the home fires burning in that way when we could not have worked outside in that storm.

I did not go back to the ranch again until the following year, 1880, at sheep shearing time. Father had built a house in the meantime, part dugout and part sod, with windows for light, something the old dugout lacked. We now had a stove to cook on, so as the men were going to be very busy with the shearing, I would come in handy to help with the cooking and I was eager for the job. Mr. Williams had a crew of ten Mexicans to shear for him that year and when they got through at his place he brought them over to do our shearing also and had to stay as interpreter, for they could speak no English and we no Mexican.

It was quite a crew to cook for, fourteen men in all, but I managed very well with a little help from brother George. He was instructing me as to quantity in the matter of meat, beans and coffee, but the bread I managed all by myself for mother had taught me to make light bread and I had brought yeast along for that purpose. So every day I made up and baked a big batch of bread. It was quite a job to knead that big pan of dough for I was not yet twelve years old and small at that.

Part 1 • Chapter 6

The boys offered to do it for me, but I would not let them, that was my job and I felt sure of my ability, for everyday that bread came out of the oven wonderfully light and flaky. Father was very proud of me but to Old Man Williams that homemade bread was something short of marvelous and he never got through praising it. He bewailed the necessity of feeding it to "those D__ Mexicans". "Can't you make up a batch that will be good and solid to feed those greasers?" he would say to me. "They never ate anything like this in all their lives and whenever you treat a Mexican 'White' you spoil him." But I just laughed and told him I could only make one kind of bread and that was good bread.

The Mexicans were a hard looking lot. Most of them had faces pitted from smallpox. One, which was half Mexican, half Indian looked especially villainous. They came on Friday and did not work on Sunday, but lay around resting, smoking and gambling. One of them had a guitar in the wagon and that evening he got it out and played and sang and then some of them danced. It was a very bright moonlit night, almost like day. One of the boys came down to the house for me and said, "The Mexicans are dancing, don't you want to come up and watch them for awhile?"

So we went up to the shearing shed and I perched myself on the top of a pile of stacked wool and watched the performance. The Mexican dance is a gliding movement with many turns and whirls and I wondered how they could do it so well on uneven ground. They finished their work on Monday afternoon and left that evening. The Indian who was sent out to get their team, which had been turned loose to graze while they were there, gave us a pretty good exhibition of bareback riding. Their horses were feeling good and were hard to drive up but that fellow could ride like an Indian all right. Guiding his running horse by the movement of his body then by the bridle reins, at a quick turn he would lay almost flat against the horse's side. On a long run he would sit bolt upright and then at the next turn he flattened along the other side of his horse. To me this was a greater diversion than the dancing had been.

Father and George started for Dodge City the next day with the huge wool sacks loaded on the wagons. Will and I were left on the ranch and it seemed very lonely after having such a crowd around. I worked and scrubbed up everything on the place and still the days seemed endless. Will was out with the herd from sunrise to sunset, part of the time I was out with him but that got tiresome too. One morning I decided I would surprise him by baking a pie, the first I ever attempted, but the pie was a dismal failure as to the crust, so my surprise was not what I had hoped to make it. Although he ate the pie and assured me it was all right, but I knew better.

Only one person came in sight during that lonely week. I had just stepped outside one hot afternoon to go to the spring for a fresh drink and there coming directly toward the house was a stranger on horseback. I waited for him to ride up but instead he pulled his horse's head a little to the south and jogged right on by. That evening when Will got in I said, "A man on horseback went by here today." Will laughed, "You scared him off. I saw that fellow, he came by and talked with me awhile and I told him to go by the house and get his dinner. He said he would, but I did not think to tell him my sister was at the ranch so when he saw you I suppose he was too shy to stop." And so no doubt the poor fellow went dinnerless that day for it was a long way to the next ranch.

I was certainly glad when Father returned and took me back up to the settlement and the home folks where times were not so dull. The following autumn Mr. Williams drove his sheep north and sold them, going out of business. When leaving he gave one of his most prized possessions to me, Old Brownie, a shepherd dog he had raised and trained. Brownie was a fine intelligent animal. He understood perfectly all commands given in either English or Mexican and obeyed them. When Brownie was out with the flock, the shepherd had little to do except to give him his orders.

Mr. Williams was very fond of him and hated to give up his dog but Brownie was growing old so his master left him with me knowing that he would have kindness and comfort in his old age. Brownie was the most sensitive creature, a few cross words would make him sad and drooping for days, a blow would have meant heartbreak for sure to him. Good old Brownie, we sincerely mourned his passing.

In the settlement along Crooked Creek changes were going on. Some new people were coming in and quite a number moving out. The Milligan family left and Father was appointed Postmaster of Belle Meade. The Post Office moved to our house. Sister Elizabeth was the assistant Postmaster and did all the work with the mail.

P.G. Reynolds of Dodge City had all the mail and stage routes out of Dodge City to points south and into Texas, a change of horses was kept at our place. The trip over our route took a week in going and coming, three days out and three days back with Sunday to lay over in Dodge City. The stage driver always stayed over night at our place both coming and going. He often carried passengers and in that way our family made the acquaintance of a number of the cattlemen on the ranches far south of us.

Part 1 • Chapter 6

The ranch owners seldom remained on the ranches during the winter months. A trusted foreman was left in charge while they spent their time at Dodge City or Kansas City or at their old home "back East." When the ranch owners returned to their ranch they would often come back on the stage. One I remember very distinctly was a Colonel Somebody or other whose ranch was in Texas. At beef shipping time in November he had gone to his old home in Kentucky and spent most of the winter there. He was a typical Kentucky colonel in appearance, tall, genial, loquacious, with a youthful form and bearing, which his silver hair belied.

His visit at the old home had been a very pleasant one and he spent the evening telling us about it. Especially enthusiastic was his description of his two young nieces, who in the interval of his several years' absence from home, had grown into young lady hood. They had acquired all the graces of the 'Southern Belles' of that period, talented musicians, good singers, and very popular, it seems. They had insisted on showing old "Uncle" the time of his life and the winter had been spent in one continuous round of dances and other social activities. It had all been very enjoyable for the time but now he was on his way back to the ranch and would be very glad to resume the frontier life that he loved and had forsaken temporarily.

As he told us in such detail of the gay, social life of these talented young nieces, my sister and I made swift mental comparisons of their lives and our own. Our life so barren of opportunity for education and the acquirement of social graces and perhaps for the first time in our lives we felt cheated, and yes, just a little envious. But the kindly old colonel never knew that for he was only trying to entertain us with something that all girls are interested in. Then rummaging in his traveling bag for a few minutes he found and presented us with a box of very delicious homemade candy the girls had insisted he take with him. This gift pleased us but it did not entirely soothe our slightly ruffled feelings.

Another ranchman that I remember well was of an entirely different type. He made several trips with the stage driver and always amused us greatly by his "Old Country" mannerisms for he had been born and reared on the Emerald Isles. A short, squat, slightly paunchy man and when coming downstairs in the morning always greeted my Mother with a cheery "Good morning Ma'am." This was accompanied by an old world curtsey and bow from the waist, a quick buckling of the knees and an almost instant inclination of the body. A tall man might have achieved this with some grace but for one with his build it was simply ludicrous, at to least to us. We found it hard indeed to keep straight faces as after greeting Mother he accorded my sister and I the same curtsey and bow.

Mischievous creatures that we were, afterward we spent hours practicing that combination curtsey and bow. However, we never achieved the smooth ease of execution of that difficult feat as one would to the manner born.

Besides all the settlers who came to our place for their mail a good many of the cowboys in the surrounding country got their mail at our office too. They would always have time for a friendly visit and a friendly chat with the family and so our friendships grew and we found the post office an interesting addition to our daily lives.

I think the most fascinating things in our experience of those early years were first the look of unexplored newness of all the vast expanse of country, then second the number of people we met. They were all new, all different; some of them perfect enigmas to be figured out, if one could. People who came from every section and walk of life, they came from all over our own country and many foreign countries besides. To one who was interested in the study of humankind, it was a wonderful opportunity.

A philosopher and writer of old once wrote, "we are a part of all whom we have met and been associated with," or words to that effect. If that were true then we of the old West are a composite lot, indeed. The families in the settlement were all, with a few odd exceptions, people much like ourselves with a common purpose, that of creating new homes in this new land. But these unattached men who wandered across our path in such numbers, they were different.

What was their interest in this new land? Certainly they were not founding homes. Free lances were what they were, for the most part traveling forth looking for adventure or fortune, hiding from the law and a lot of them just looking for the fun they could find in the carefree life of the cowboys of those days. At that time the Code of the West was partly, "Do not ask a stranger his name or his business or where he was from. Do not manifest any curiosity as to his going and comings."

So all these strangers that we were brought in contact with at our house were received with a friendly courtesy and an open hospitality, no matter what their appearance. And more often than not, this was the very thing that brought a response in confidence and an intimate glimpse into the lives of some of these men. But not always was this so.

On one of his trips to Dodge City, Father brought back with him a man to work on the ranch for a couple of months. Our "Mystery Man" was what we called him. He gave his name as Frank Carter though we never believed that was his real name. He had black piercing eyes, blue-black hair and a poker face. He did not look like a man that had been living and working in the open air and his appearance was such that we all distrusted him. Father said he had probably made a mistake in hiring him but he seemed anxious for the job so thought he would try him out.

To our surprise he proved to be a very dependable helper on the ranch and the boys liked him very much. But they noticed that he never told anything about himself or his past and he never got any mail. He seemed quite content to live in that isolated place and herd sheep. At the end of the time he was hired for, instead of leaving he went over on Remuda Creek, about three miles east of the sheep ranch, and made him a tiny dugout in a side draw, an ideal hideout. There he lived the entire winter seeing no one except my brothers. One day when they dropped in to see him he told them he had narrowly escaped death there all by himself. While preparing coyote bait with strychnine, he had carelessly left a bit of it on the point of his knife and then cut some tobacco, so getting a dose himself. He had been terribly ill for a couple of days, but had pulled through. When spring came he disappeared as if the earth had swallowed him. He made his way out on foot, without ever being seen by anyone, and to this day remains a man of mystery, but there was a good many such as he in those days.

Cimarron Chronicles

Chapter VII

The matter of schooling was a great problem for our parents the first year on the homestead, although I think it troubled their children not at all. So, when autumn came, it was decided we must study at home. My brother, Will, had had the advantage of several courses in eastern Iowa Normal School, and rather unwillingly he now undertook the job of tutoring us. We had lessons to get during the day and we recited them to him in the evenings.

By the next winter, though, he was at the sheep ranch, so a small frame schoolhouse was built close by the Peter Blair home. We had a three or four-month term of school, the teacher was a young lady of the settlement. Her family had just recently moved in from New York state, they were educated, refined people. As there were four or five grown young folks in the family, all musicians and good singers, they proved quite an asset to our social doings. But poor Alice, our teacher, was painfully shy and quite bewildered by the way of the West.

She always blushed when she had to replenish the fire with "cow chips," the only fuel we had and she confided to us girls that it was "dreadful." Then just when she was getting used to that, something more dreadful happened to her, for one morning she had a new pupil. He was thirty-five years old, a bachelor, and had just recently arrived from Norway. He could barely speak a few words of broken English but was anxious to learn to read and write. And so he came with slate and primer and squeezed his man's bulk into one of the little homemade desk seats and waited for his lesson.

Our teacher looked as if she would like to sink through the floor. She paid no attention to him and we went on with recitations as usual until recess time. The younger children all went out to play but Cousin Mary and I, the oldest girls in school, remained in our seats. The new "pupil" also remained seated so our teacher came and sat by us pretending to help us with our homework and speaking very low she asked, "What am I supposed to do with HIM?" And we answered, "Teach him just as you are doing with the other pupils."

So when school was called, with her face flushed a deep red, she began the task of teaching him to read. It was very diverting for the rest of us, for she never got over blushing when hearing him recite and I am sure she was very glad when the term ended. Her pay was fifteen dollars a month and board since this was a subscription school. As there was no county nor was there any school organization as yet and of course, no teacher's certificate was required.

Although our school facilities were limited Father impressed upon us that we must not grow up in ignorance. He saw to it that we kept up our studies at home during all the long winter evenings. After our family had moved to the ranch we were isolated indeed, especially the first year or two, for we had no neighbors at all.

It was in 1882 that we moved to the ranch. The main reason for abandoning the homestead was water. In all the numerous wells, that the men had dug in different areas of the place, the water was unfit to drink. Years later we found that the whole neighborhood was underlain with the most wonderful artesian water but it was too late for us.

It was quite a change for us, leaving the settlement and all the young folks, especially our cousins who lived so nearby that we were together almost constantly. The Post Office was moved to Uncle John's and it was this which perhaps we missed most of all. For in one way or another we girls had managed to get a lot of fun out of the Post Office and the people it brought to our place.

Yet, as I look back on those first years at the ranch and recall some of the most tender memories of my girlhood days, the family was all together again. We were thrown together completely on our own resources for amusement and pastime. We all had work to do but there was time for play too and our greatest fun was to think up practical jokes to play on each other. In the evenings, after lessons were over, Father sometimes told us stories of his boyhood in Switzerland. We all loved to hear them. Sometimes we played games, such as cards, dominoes, chess and other games but more often it was books that held our attention. We read, discussed and fairly lived in some of them.

A friend gave us all of Dickens's works, paper bound, and I read most of them aloud to the family in the long winter evenings. Dickens was a bit slow in his way of unfolding the plot of his stories, but his character delineation was so rich and complete that they were real folks to us and how we did enjoy some of them.

Often for two or three weeks at a time we would see no one outside of the family. It was during such dull times that some one invariably would pull off the "time worn trick" of slipping unnoticed outside and knocking loudly on the door. When some member of the family with an expectant look in his or her face opened the door the offender had to dodge. Then sometimes some one would come in and announce, "I see a couple of horsemen coming." When, of course, there was no one in sight, but the trickster got paid back in his own coin with interest.

On rare occasions during the fall and winter, we were enlivened by a freighter coming by and staying over night with us. They were always equipped for camping, but Father would have them come in and stay with us. We were glad of the chance to hear outside news, the freighter equally grateful for warmth, food and companionship, and this usually proved most entertaining. These men were freighting food, grain and supplies to the ranches south of us.

The winter of 1883-1884, I stayed with our friends, the Blair family, and again I went to school in the little frame schoolhouse nearby. Not a long term, four or five months at the most, but I got a lot of good out of it for we had a good teacher. She was a young lady from Chase County, Kansas, with several years of experience in teaching, and was willing and eager to give us all the help she could and I made the most of the opportunity.

Father brought "Nellie" up for me to ride for all the young folks had riding ponies and we went everywhere on horseback. It was a lot of fun and we thought nothing of riding to Church up at the Emerson Grove on Sundays, a little matter of twelve miles, and back in time for our own Sunday School in the afternoon; or of going to a party up beyond "Three Bends" where Fowler, Kansas now is.

The wide Jones and Plummer Trail was always smooth as a floor and hard as pavement and when we got on it there was usually a free for all horse race. One bright moonlight night four couples of us returning from a social gathering at something past midnight filed down the trail. At the Three Bends there lived an eccentric old bachelor in a filthy shack he called a roadhouse for the accommodation of freighters. As the first couple reached this place Cousin Ed yelled, "Hey, Hensley. Wake up, supper for eight." A sleepy grunt for reply was heard and they rode on. When the last couple rode by, there was a light in the shack and a great rattling of pots and pans. Hensley was getting supper.

We had a great time that winter, often the crowd from Three Bends came down to Uncle's to spend the evening. Always when the crowd got together there was music and singing, and "charades" were very popular with some of the youngsters. But I missed the home folks and whenever possible on week ends Cousin Mary and I went down to the ranch. One evening when the family was not expecting us and it was after dark before we got home, we planned to pull off the time worn stunt of knocking on the door and fooling the folks.

We sometimes wore spurs when riding long distances so now we rode up near the door, dismounted, and Mary stood back in the shadow holding the horse. I walked up to the door with spurs jingling and clattering and knocked. As we hoped Elizabeth came to the door and stepping back out of sight from the open door with more spur jingling I spoke in a hoarse masculine voice, "Good evening Ma'am, can ya all keep a couple of strangers overnight?" "Why, yes, I think so," she answered innocently. That was too much for us, we whooped. And then did we get a pounding from that indignant young lady. I'll say we did!

That winter the cattle were thick all over the country, the big flat between Uncle John's and our ranch on Crooked Creek was covered with grazing long horned cattle of many different brands. Sometimes when crossing the flat, the cattle would all be headed north, walking as they grazed with every head turned to the north. At such times even though it was warm and cloudless at the time, we knew that a "Norther" was brewing and that the animals through some strange instinct sensed this and would go to meet it until the storm struck and then would all turn tail and drift with the wind to the south until the storm was spent.

Most of these thousands of cattle had been turned loose on the Arkansas River in the fall, their owners never looking after them until the spring round-up when many of them might be found on the Canadian River in Texas. Another outstanding memory of that winter was the amount of fame the Blair boys and a young man named Will Siebenthaler brought in.

Bill, as we called him, was a young homesteader neighbor of the Blairs, boarding with them that winter while making some building improvements on his place. He was a fine marksman with a rifle and shotgun, and the lagoon just half a mile west of the house was full of ducks and geese. The boys would slip down there after school was out, and come back in time for milking and evening chores, loaded down with game. Then after supper all hands picked and dressed the birds.

I think Mrs. Blair stored enough feathers in her attic that season to make two feather beds. Some of us got so tired of duck and goose that we could hardly eat it any longer, but these boys never tired of the game. We served them cooked in every way we could think of; baked, boiled and stewed they appeared on the table twice daily.

Part 1 • Chapter 7

That same season in late November, Father took Bill to the ranch on a hunting trip. They had grand sport and good luck getting a number of both deer and antelope. Father salted and smoked some of the meat and it made a welcome variation from the ever-present mutton. The deer always stayed in the rough country but the antelope always grazed in the flat open places.

On our trips across the flat going to or from the ranch we invariably came across a band of antelope. Their curiosity at the sight of persons on horseback always allowed one to ride up close before they turned and with a clashing of antlered heads dashed away as we pursued them with wild yells. But in no time at all they always left us far behind as they skimmed away in seemingly effortless flight until the gray brown of their bodies blended with the gray brown of the winter prairie and only their short, fan like tails, gleamed snowy on the far horizon.

Soon after the family moved to the ranch a cattle ranch was established on the creek about two miles farther down, the "A H Bar" outfit. This seemed quite a calamity to have two ranches as close together. At first to us it seemed that the attitude of some of the cowboys was just a little overbearing toward us of the sheep ranch, but Father said the way to get along was to be friendly. It was not long until we were acquainted with all the boys and seldom a day passed that one or more of them did not stop in for a little visit with the family.

There was "Peggy" the cook, a pleasant faced young man with a wooden peg strapped on at the knee that served in place of an artificial limb. The boys claimed that it came in very handy to stir up the fire when Peggy was getting meals over the fireplace, but he limped about very cheerfully and never minded the banter of the boys.

Win Hobbs, one of the cowboys, a slender young chap, was almost dandified in his manner of dressing. He was a long solemn looking face and seldom smiled, but behind that solemn visage was a disposition for fun, mischief, devilment that I have never seen surpassed. And it was not long ere "Kid" Hobbs was known far and wide over all the ranges for his dare devil tricks and pranks. There were few dull times when "Kid" was around.

The foreman of the A H Bar was Thomas Johnston, better know as "Slick," and in the next few years his visits in our home were notably frequent.

In those first years of our life on the sheep ranch, the coming of spring was usually heralded by a three-day rain, often for three days and nights the skies were leaden and rain fell without ceasing.

Days and nights of utter discomfort for men and beast for it invariably befell that the cowboys were just starting south for the long spring round-up and so got a thorough soaking. While we, of the ranch, had troubles of our own for the herd had to be taken out to graze no matter what the weather. With wet fleece and chilled bodies the sheep ran and scattered until the herder became almost frantic with exasperation and fatigue and it generally took two or three of us to get them gathered up and corralled.

In the meantime the house roof developed a surprising number of leaky spots. Pans and pails were set about here, there and everywhere to catch the drip. At night we were often awakened by that drip, drip on head or face or coverlet when we hastily arose and pulled the bed down or out, as the case might be, only to be awakened later on by a new leak. Cooking was a task for the fuel all got wet or at least damp.

On the fourth morning the rain ceased and by 10 o'clock the sun shown out warm and bright, meadow larks sang joyously, the herd settled down to peaceful grazing, all the bedding and wet garments were hung out to dry and "joy reigned supreme." A little earlier than the usual rainy season, our horses always took a notion to visit the Cimarron River. There were no fences anywhere of course, but we always kept one horse hobbled so he could be easily caught. But some mornings we would wake up to the fact that all the horses were gone and we were left afoot. The green grass came earlier on the Cimarron River bottoms than it did on Crooked Creek and the horses sensing this had pulled out. Then one of the boys with a bridle over his arm started south. Sometimes the hobbled one might be found far down in the hills, but I know once or twice they had to walk to the river were the scamps were found and the way they were brought home was not slow.

Early in May the lambs arrived and then for about a month everyone was kept on a jump. The wethers were cut out and kept in a separate herd ranged farthest from home and penned separately at night. The ewe herd was kept closer home and every evening one or two of us went out to bring in the ewes with lambs that came during the day. Then these soon made a third herd that was kept very close to the home place. Father and one of us girls usually watched them for they gave the most trouble. A bunch of young lambs at play are the sweetest, most innocent little creatures on earth and one never tires of watching them as they skip about in rollicking play. But a young lamb awakened from a nap is nothing short of fiendish.

The dear little innocents require a great amount of sleep and while they are resting peacefully the ewe mamma's graze away from them and keep on going, it is then the business of the herder to stir up the lambkins to make them keep up with the herd. That is all right if there is a bunch of them together but just arouse one by itself asleep under a sagebrush and see what you get for it. Lambie jumps up, turns his back on the herd and sprints wildly in the opposite direction, never by any chance does he run toward the herd.

Everyday we ran miles and miles chasing them, sometimes we would catch them or turn them in a wide circle when they eventually blundered into the herd. But some of them we lost entirely, we might run across them in a day or two, weak and nearly starved, by that time the mother would not claim her offspring, however, generally the coyotes got them.

All during the lambing season every day brought an incredible amount of footwork for us all. So that when evening came and the various herds were penned and the last lamb was in the fold we were all weary beyond words.

With June came shearing time, then Father and the boys helped with the shearing and we girls herded. The different flocks were then all thrown into one herd. The fleeces were packed into huge wool sacks that when filled weighed five or six hundred pounds each. They were later freighted to Dodge City and sold or shipped to Kansas City.

It was at one of these times, when the men all busy with work at the corrals, that my sister and I were holding the flock about a half a mile from home. She was on one side of the herd and I on the other. As we waited and watched the sheep grazed quietly, I had been sitting on the grass "Turk" fashion until I got tired of that position and extended my feet. With my arms thrown back of my head, I let myself fall on my back for a rest. With one bound I was on my feet with a yell that would have done credit to an Indian on the warpath for I had plumped myself on a little clump of prickly pear cactus which was hidden in the grass. One of the prickly pears was snugly embedded in my back right between the shoulder blades where I could not remove it.

In an agonized voice I called for my sister. "Come here quick, run," I yelled and ran she did. When she got in speaking distance, she gasped, "What is it?" "It's a cactus," I replied, "right between my shoulders where I can't reach it. Take it out quick." She stopped dead in her tracks, wrath overspreading her countenance. "You miserable imp," she stormed, "yelling like that and scaring me half to death just for a measly little cactus in your back. I thought it was a rattlesnake bite the way you yelled. And just for that I think I will leave that cactus there just for punishment." And I knew she was quite capable of doing that. "You ought to be glad it was not a rattlesnake bite," I answered meekly, "please take it out." This cooled her wrath so the cactus was removed though not too gently.

With the lambing season and the shearing over, the long lazy days of summer followed and then there was time for a visit with our cousins in the settlement. Sister and I usually went together for it was a fifteen-mile or more ride and no houses in between. Sometimes we passed herds of cattle being driven to Dodge City, and always on the Jones and Plummer Trail we passed ox trains carrying freight to points far south such as Mobestie, Tascosa, Fort Elliott and places between. They were great heavy wagons each with its canvas covering.

There was always a lead wagon with one, two or more trail wagons attached and hitched to the lead wagon. There were anywhere from six to twelve yoke of oxen, the driver always walking at the left of the string of oxen, carrying his huge bullwhip, every crack of which was like a pistol shot. Sometimes there was as many as forty wagons in a train, it was an imposing sight. I shall never forget the slow swing of the heads of oxen, each under its heavy yoke, slowly and patiently pacing off the weary miles. The drivers were rough looking men, many with swarthy Mexican faces, and we always passed these trains at a swift canter.

Chapter VIII

Our visits to the settlement were very enjoyable, but I think even more so were the times when one or more of the girl cousins came home with us to spend several days or a week.

Cousin Mary, sister Elizabeth and I were devoted chums. We shared all our secrets, hopes, fears, ambitions and love affairs for by then we were reaching young womanhood in a land where men were so plentiful and girls so scarce. We did not lack for attention, so when we three got together we had gala times and there was so much talking that we scarcely took time for sleep. When times were dull we planned something mischievous and always for diversion there were long horse rides.

Remuda Creek, a few miles east of us, was a favorite haunt for it was an enticing spot in spring and summer with its many fine old cottonwood trees making shady retreats. It had hundreds of plum thickets that in season were always red with luscious ripe fruit. Tall grasses, high rushes, deep canyons, on either side of the creek bed, made a place of wild beauty that never failed to appeal to us. Invariably, too, we rode up to the old abandoned Carter dugout and with delicious thrills of fear peered within. Perhaps the outlaw had returned and was once more in hiding there or perchance a bobcat had taken up its abode therein, such surrounding one might expect to fine anything. It was in this mood we had been riding all afternoon, through all the "spooky" places, along Remuda, not seeing a soul nor indeed expecting to. When starting for home near sundown, at the mouth of the creek coming up out of it suddenly on to a little flat, we were surprised and a little startled to see a large herd of horses grazing, a campfire and a number of men lounging near it.

As we drew near several of the men hastily mounted their saddle horses and rode out to the herd. One man lying on the ground, by the fire, turned over and hid his face in his arms. But one man stepped boldly out and came to meet us. A cowboy we had known for several years, he had worked on nearby ranches and had been at our house many times.

"Good evening, ladies," he greeted us. "Out for a ride?" And we chatted on for a few minutes in casual talk, although we were consumed with curiosity. Who are these men, where have you been, and what are you doing with all those horses? These were some of the questions we were eager to ask but the Code forbade. Our acquaintance appeared ill at ease and when we said we must be going on he suddenly laid a hand on sister's bridle rein and said, "Wait, I want you to all promise me that you will never tell anyone of seeing this outfit. Forget that you ever saw anyone here." He was in such deadly earnest that we promised and that promise was kept.

Of course, we talked it over and drew our own conclusions, for a time we half expected to hear something further about the mysterious outfit. However, we never did and so in time we never even spoke of it to each other. I doubt that any of us ever "forgot," but we never asked any questions of anyone. In very recent years, when all the principals in this little drama but myself had passed to the great beyond, I told the story to an old timer and so got the answer to some of the questions that the Code had forbade us to ask.

Those horses, he told me, had been stolen in Montana and driven to Texas. The horses were being handled by a "ring" of horse thieves, each change of men being familiar with the country through which the stolen band was being conducted. As to the identity of the other men with the herd I am still in the dark but it was quite obvious that some of them were afraid of being recognized.

But this episode was merely an adventure. An experience my sister and I had soon after sank deeper into our consciousness. Father and Mother had gone to Dodge City to be away for three days and two nights. Only Brother Will was at home just then caring for the sheep. The corrals had been about a quarter of a mile from the house and a little shelter with a bed nearby. This is where he slept every night for coyotes had been disturbing the flock at night. So that left my grown sister and myself alone to look after the three younger children. It was the first night we were alone that Brother Arthur, then a small boy, had an attack of croup so severe we thought he would die.

We were awakened about eleven o'clock by his hoarse strangling cough. Dressing hurriedly I made up a fire while sister wrapped him in a blanket and held him in her arms. We applied one remedy after another but he grew steadily worse. I remember running out to the old dugout to get something and coming back, while still some distance from the house, I could hear his labored breathing. I had heard animals breathe like that when being choked down with a rope.

Part 1 • Chapter 8

Heavens, how frightened I was. There was no time to go for Will, the others were too little to send. We decided that our only hope would be to induce vomit and this, after repeated attempts, we finally accomplished by a rather severe method. After throwing up the thick ropy stuff that was strangling him, he was soon breathing easier and after a time fell into an exhausted slumber. But there was no sleep for us, we watched him carefully the night through and all the next day fearing another attack. But we all slept peacefully the second night and when our parents returned the following evening all was well at home.

In those days when the nearest doctor was at Dodge City, some sixty miles away, and transportation was so slow as compared with the present time, one simply had to depend on their wits to do the best they could and trust the rest to Divine Providence.

At a later date we suffered much anxiety when my youngest brother Elmer, then a little fellow of three years of age, was bitten on the hand by one of the large red ants that had a nest near the house. We thought little of it at first but after a time the hand began to swell and kept on swelling until it was perfectly round, then the arm puffed up clear to the elbow. We did everything we knew to do, but it steadily grew worse. Father had gone to Dodge City and Mother was almost sick with worry. We decided we must have help. The Martz Family had moved in on the creek about three miles west of us. We were not very well acquainted with them but Will rode up there and told Grandma Martz about our troubles and she, a true pioneer with much experience, came to our rescue. With her little granddaughter behind the saddle, she rode down to our place and stayed all day and night, a perfect Godsend to our troubled family.

Anyone who has not been through such an experience can never know what a relief it is to have someone like her to come in and take charge. She was so cheerful and so sure, "Why," she said, "That will be all right, just use plenty of baking soda poultice on it, that will draw the poison out." And it did, for by the next morning the swelling was gone, leaving the little hand all wrinkled and then Grandma rode back home leaving us all cheered up and grateful.

Almost every pioneer community in the old days had its "Ministering Angel," a person who went wherever there was sickness and sorrow and did what she could. Grandma Martz was just that to the widely separated ranch homes around us.

The Martz family left Missouri in 1879, and traveled by wagon, driving with them some 100 head of cattle. They had gone west into the edge of Colorado where they found the drought so severe that there was no grass for the livestock, they then turned back settling far up on the Cimarron River. But in 1881, they moved near us on Crooked Creek, building a long low sod house of three large rooms with a porch extending the entire length of the south side that faced the much used Jones and Plummer Trail where it crossed the Creek. This soon became a popular stopping place for cattle ranchers and cowboys traveling the trail. The family consisted of Grandpa and Grandma Martz, their son Jim, his wife and several small children and an adopted daughter, Miss Bell Sewall, a young woman of fifteen or sixteen whom soon became known as the "Belle of Crooked Creek." For several years the Martz family was the only one living near us.

On Stump Arroyo several miles northwest of the Martz place was the ranch home of Labon Lemert and his family. Mr. Lemert settled at his place in 1879, and we made his acquaintance while we were living in the tents on the homesteads. He came to our place one day on foot very much excited and worried for his team. He said that they had been stolen or had strayed away and he begged for a horse to ride that he might more readily find some trace of them. As we had nothing but the mules my Uncle John said he might have one of them to ride and offered to accompany him. So they started off in search of the missing horses and in a few hours were back again. They had found them several miles further on, northeast along the Jones and Plummer Trail, evidently headed back for the eastern Kansas home they had come from.

Mr. Lemert was very much relieved, returning to the home he was making for his family who arrived some time later. His older children, Bernard and Irma were among our young friends when we lived on the sheep ranch later.

The Edward Boyer family, some ten or twelve miles up the creek from us, was another of our neighbors in those days of widely scattered homes. They moved in there in 1880 or 1881, with a small herd of cattle and went through a good many hardships in building up the ranch home. The older children, Mada, who was my age, and Richard, several years younger, were able to help some but there were three younger ones, mere babes. The cattle they owned could not be turned loose on the open range and had to be constantly close herded. A job that young Richard undertook, although he was so small one wondered how he managed to stick on a horse's back. The family had not been long in their new home when sickness and tragedy fell to their lot. The father was away from home for some days. Whether on one of the long trips to town or off on the ranges looking for strayed cattle, I do not know, but the mother was alone with the young children when they

Part 1 • Chapter 8

became terribly ill of some malady that was at first supposed to be a virulent form of small pox but later proved to be black measles.

There was no one to send for help, but one day a stranger, a cowboy, came by and when told of their need, he went out and found the father who returned home in the shortest time possible. Two of the children had died, the baby boy and a lovely curly haired little girl of three. Here again Grandma Martz came in and stayed as long as needed, giving help and comfort to the sorely afflicted family. Later when new babies came to the Lemert and Boyer homes it was Grandma Martz who was both doctor and nurse on those occasions.

It was these three ranch families on Crooked Creek and the N.J. Rhodes family on the Cimarron River who were our associates in those first few years on the ranch, although, of course, there were always plenty of cowboys coming and going. Mr. Rhodes with his oldest son Wiley moved their cattle to the Cimarron in 1878, when the R.K. Perry Ranch was established. A few years later the Rhodes family came from their former home on the Pawnee, in Kansas, to live on the ranch and the three girls, May, Clara and Hattie, proved an attractive addition to the ranch personnel.

In those days when we went visiting to one or the other of these ranch homes it was on horseback. We would spend the day, going early in the morning and returning home in the evening. Those long rides were a part of the day's enjoyment for horseback riding was one of our chief pleasures, although I sometimes wonder now how we could endure those long riding skirts we wore in those days and the sidesaddles that are now only seen in Museums.

But we were proud of those riding skirts and the longer they were the better we liked them. So do time and customs change. The "Belle of Crooked Creek" soon was married to one of the young cowboys, Mr. Dave Mackey. Their marriage at the ranch home of the Martz's was a social event.

Miss May Rhodes, too, became the bride of another one of the cowboys, Mr. Frank Murphy, in a rather unusual ceremony. The wedding was performed on the open prairie in a snowstorm. The wedding party proceeded from the ranch home in "No Man's Land" to a point just over the line in Kansas where the license had been issued. As "No man's Land" in that time had no organization of any sort, it was feared that a ceremony in that forgotten land might not be legal.

Chapter IX

The day, usually in late May, when the Spring Round-up came down Crooked Creek and past our place, was always an interesting event for us. For a month or more our cowboy friends had been off to the south in the general round-up and we had missed their coming and goings. Now they were back and they were coming across from the Cimarron and down Crooked Creek where a round-up was held on the creek somewhere below us. First came the circle riders in bunches of two or three, shoving all cattle from the hills and valleys, to the converging point where the round-up was to be held. Later came the chuck wagons, the horse herds and the day herds of cattle being taken to home ranges.

After weeks of placid uneventful days, the life and movement of the round-up was quite a diversion as we watched them moving by, down along the creek, but we seldom saw any of the boys then, for when a round-up was on there was work to do and no time for visiting. Besides after weeks of round-up life, they were bearded and unkempt and had no desire to present themselves to feminine eyes.

But round-up over and back to the ranch for a short stay, they soon presented themselves shaven and shorn and dressed in clean "boiled" shirts. This is when we heard of the main happenings, riders racing in the storm and darkness, horses falling, riders hurt, the lightning playing a continual stream of fire over the cattle, horses and men. We heard of fierce spring storms when the herds stampeded and these rather grim details always interspersed with the lighter vein—whose horse won most of the races that season, and various side splitting jokes and stunts pulled off on each other.

After all was said and done, we got the impression that in spite of hard work and exposure to all kinds of weather, the boys were having the time of their lives. They would not have traded jobs with anyone and only wished that this fascinating adventurous life might go on and on, indefinitely, without interference from the "grangers" who began in 1884, to put in their appearance in ever increasing numbers.

Several families moved in on the creek beyond the A.H. Bar Ranch, and quite a number homesteaded near by us. For the first year or more they were busy building and making other improvements and little farming was attempted so as yet the cattle roamed unmolested. But for the cattle ranchers "the handwriting on the wall" was becoming apparent even then.

One of the things that troubled us on our sheep ranch during the years from 1883-1885, was the increasing herds of range cattle that came drifting in on us from the north in autumn and early winter. In order to preserve range for our flock we were forced to shove these cattle on to the Cimarron River to get rid of them. These were not cattle belonging to ranchers near us but were mainly herds turned loose on the Arkansas River and beyond which were allowed to drift where they would.

Early in the spring of 1884, new settlers began to come in and they continued to come all that summer and autumn. A perfect rush of them, seldom a day passed without a party of home seekers coming by or stopping at our place. They came in wagons, on foot, and in livery rigs, all excited and asking numberless questions about the land, the country, the rainfall, and such things. It was not long before we had neighbors galore. It was quite a change for us and I think we children especially enjoyed the excitement of meeting so many new people, some of them so very odd but most of them, of course, just nice ordinary every day people.

Our home was always full for many came to stop overnight, for dinner or for a few days until some little work could be done on the new claim. A surveyor was brought in to establish lines and corner stones. He made our place his headquarters.

I often wondered how we managed with such limited houseroom to furnish food and shelter for so many, besides our own family. But in those days there was always room for all comers and we just did the best we could. No one complained even though the improvised extra beds made down on the kitchen floor for the men must have been very hard indeed, but these new comers were not used to sleeping in the open on the ground and appreciated a roof over their heads.

Soon we had near neighbors, quite a number of them. Bachelors to be sure, but there were also a good many families, the nearest only a half a mile away. How mother and we girls appreciated walking that short distance to spend an hour or two, or perhaps the entire afternoon with the friendly little mother in that soddy home.

Part 1 • Chapter 9

Mr. and Mrs. Wurdeman lived there with their family of several young children. After a few years they left, moving to eastern Kansas but she now lives in Meade, Kansas, where she has children, grandchildren and great grandchildren for this cheerful and well preserved pioneer is now well past eighty years of age.

Soon we were quite surrounded by a community of German Lutheran people from Missouri, many of whom came with very little worldly goods, but by thrift, economy and the ability to get along somehow on very little, almost all of these became permanent settlers. The older heads of families have all passed on but their children and grandchildren remain in the country.

While all of the first settlers made their homes on or near streams because of the need of water and the depth of dug wells on the high flats, now with the rush of homesteaders in 1884 and 1885, all of the high land too was taken. There was more rainfall during those seasons and all the buffalo wallows held rainwater, many shallow lakes were everywhere and over the flats.

The summer of 1885, there were no hot winds, the weather was unusually cool and cloudy so the evaporation was not great and the newcomers could not be made to believe these were not normal conditions. When the older residents assured them these lakes and ponds would soon dry up, many of these land seekers would reply, "Ah, you ranch fellows just don't want to see us farmers come in here." Some were even less tactful in their replies plainly insinuating that their informants were lying.

So, as my Uncle put it, "It's no use trying to tell the fools anything, they will have to find out for themselves." And many did so to their sorrow. The ones who came with little had little to lose. But for many, a family sold a good, well improved farm home in the East and in two or three years of "improving" this homestead and having living expenses, without a cent of income and the return of drought conditions, soon woke up to the fact that their money had been wasted with no possibility of any return.

I have in mind one example of this. There was a family who moved in on the flat north of us. We passed rather close by their home on our way to the new town of Meade, just being built and springing up in mushroom like growth. These people built, or had built, a large well-made sod house, nicely plastered and finished on the inside. They had shipped out much fine furniture, lace curtains hung at the windows, a canary sang in its cage in the sunshine at the deep window, other improvements on the place were in keeping. These people had confidentially expected from the good black soil of the homestead a quick response in fine crops that would enable them to continue to live in the manner and style they had been accustomed to in their Eastern home.

Almost before they were aware of it, their money was all gone, there were no crops, and the wind blew incessantly as it got dryer and still dryer. The woman grew to hate this country; she shut herself into her home and would not mingle with her neighbors. When visitors rode up to their place they were not asked to dismount and come in.

Worried at the loss of all their resources, too proud to acknowledge defeat and go back home "broke." The climax of tragedy for this family was reached when one day it became necessary to commit her to an institution for the insane.

Father and Mother went down the creek a couple of miles to visit some neighbors. The three younger children were at home as some other youngsters had come to spend the day with them. At that time of the year it was our custom to range the sheep on the south side of the creek and save the northern hills for winter grazing. So I had the flock not much more that a half a mile from home but across the creek. It was a very hot day and while the sheep took their noonday "siesta" I rode to the house for some lunch and stayed with the children for a while. Soon afternoon clouds began to pile up in the northwest and I kept watching anxiously, but they hung off there for hours. Then very suddenly there was a greenish loop about those clouds and they seemed to move closer. I gathered the sheep up at once and started them toward home but the clouds moved faster than I could move my stubborn charges. Always sheep are hard to get across a stream, no matter how shallow, and our flock, although so accustomed to the creek, were particularly exasperating when a storm was brewing.

I had them down on the creek bank and was working frantically trying to get them to take to the water when the storm broke. It was impossible to move them another step. I was right by the home of one of our German neighbors, the Gruenken family. They had a small dugout in the bank a short distance from the creek. Mr. Gruenken, his wife and small son had also gone visiting to a neighbor on the north side of the creek, but had left the older children at home, a boy of ten and a girl about seven. I took refuge in their dugout with the children. The dugout faced the creek and I stood in the doorway watching the sheep and thinking that the minute it stopped raining I would get those "critters" across or know the reason why. But the rain fell in torrents for what seemed to me an endless while, then it checked a little and donning my slicker I went out to the herd, moved them a few feet and then the bottom just seemed to fall out of the sky.

Part 1 • Chapter 9

Such a deluge I had never seen before nor have I since then seen its like. The rain fell in such a blinding sheet that I could not see my hand before my face and it is actual fact, I could not breathe. I had a panicky feeling of being smothered and made my way back into the dugout where I then stood broom in hand sweeping out the water that was falling too fast to be carried off by the slight incline of the passageway. The children said nothing but looked terrified.

When the rain practically ceased so my flock became visible once more, I saw to my horror and consternation that the creek was rising rapidly. The sheep had moved up onto the highest part of the bank and the rising water had closed in behind them, leaving them on a little island with swirling water ever rising higher and higher for already the stream was out of its banks. While I stood there helpless, in utter despair, a horseman galloped up. It was Lee Miner, one of the A.H. Bar cowboys. "Your father sent me to help you," he said. "I saw him on his way home and he knew you would be having trouble and as I could get here sooner than he could, he sent me on ahead. But I don't know what we can do now for the creek swam my horse when I crossed it just now."

While we stood there debating on what could be done Father rode up on the opposite side of the creek. He had reached home, jumped on a horse bareback and came full speed. When he reached the creek bank, the horse stopped dead still and refused to go in. Time and again he lashed that animal to the very edge, but each time the horse reared and wheeled away from that raging torrent.

I was thankful that the horse refused, for Father could not swim and I was sure he would be drowned. Finally he had to give up and move farther back for just then another big wave of water came and while we stood there the island slowly disappeared and the sheep were swept away like much chaff. That was the darkest hour of my young life when I saw our flock of 1200 head of sheep, our only means of livelihood, being carried away in that awful flood.

Lee hopped on his horse, "Stay here," he said, "maybe I can pull some of them out as they float against the bank." And he dashed away. In no time at all the last one of them had swirled out of sight around a bend. I was too discouraged even to get on my horse and follow. I was sure every one of them would drown and anyway, what good would it do to pull out a few along the bank. A few, we might as well have none. Besides it was getting dark now, soon it would be night. I was too miserable even to cry. Only one consoling thought I had. Father had not drowned.

However, a short distance around that bend was a wide sandy beach on the south side of the creek and here the main bunch of the sheep were eddied out into safety. Lee on the south side and Father on the north followed on down the stream for some distance pulling out sheep as they floated against the bank, saving quite a number in that way.

Just as night fell, Lee came back. We gathered up our flock and bedded them down on a little grassy flat just back of the dugout, for it was quite apparent that we would have to spend the night there with the children. The clouds had all passed over and soon the sky was brilliant with myriads of stars.

We staked out our ponies and went into the dugout. There was a tiny cook stove and a little dry fuel behind it, so we made a fire to dry our feet. Our slickers had kept our bodies dry, but our feet were soaked, the fire felt good. The children, no doubt, were glad to have company for the night as their parents could not get home, but they were shy and could scarcely speak any English at all. After eating a little cold lunch, they climbed up on the high bunk bed and soon were asleep.

Lee made some coffee after awhile, but I was still feeling too "low" to care for either food or drink. While he sipped his coffee, we sat on opposite sides of the little cook stove with our feet in the oven and "talked of many things, of ships and sheep and sealing wax, of cabbages and of Kings."

Occasionally we would hear coyotes howling and Lee would go out to see if they were disturbing the flock. The moon came up about midnight and it was almost as light as day outside. The coyotes were only rejoicing over the tasty mutton they were finding strewn along the creek bottoms all ready to be feasted upon. And so the night dragged on. The children slept fitfully tossing, moaning, and talking in their sleep. We soon found the reason for their restlessness for driven in by the rain, no doubt, the dirt floor of the dugout was literally alive with sand fleas, the hungriest, most blood-thirsty brood imaginable. They almost ate us up before morning and we kept our feet in the oven all night just to keep them off the floor. Lee with his high boots fared somewhat better that I did, but we were glad when the short summer night was over.

The creek was still rolling bank full but by about ten o'clock that morning Father came across and the three of us forced the sheep across and took them home to the corral where they were counted out and the loss was found to be between 250-300 head, much less than we had expected. Mother had a good hot meal ready for us when we returned which was much appreciated and so after all, this unpleasant

Part 1 • Chapter 9

experience was not the tragedy I had pictured it was going to be for by the next spring the lamb crop more than made up for the number lost.

The boy and girl who shared their dugout home with us that night are now grandparents. Only recently I met this girl and together we talked over the rough experiences of that dreadful storm of long ago. But the storms that lingered in the memories of all in those olden days were the blizzards of the winters of 1885 and 1886.

My outstanding memory of the first of these was not of the storm itself for our house was warm and we never suffered from cold in it, no matter how severe the weather. The blizzards had lasted for a day and night and the next morning dawned clear but oh, so cold.

Snow and ice covered our world. About two o'clock of that afternoon one of the boys came in from outside and said, "I hear such a strange noise outside, come on out and see what you can make of it." We stepped out and stood in the icy wind listening. From the north there came a low, deep moaning sound and while we stood puzzled, staring in that direction, around the foot of the hill and across the little meadow that lay east of our house came a mass of freezing, starving mass of cattle. There were thousands of them in a close massed bunch. In the lead were the big three and four year old steers, then the cows, many of them with calves by their sides, after them came the smaller yearlings and "doggies."

For an hour they filed by and then for another hour came stragglers too weak to keep up with the main bunch. A few of them sought the shelter of our buildings and died there miserable.

These cattle had come from the Arkansas River and beyond. It had become customary for cattlemen to buy a herd at Dodge City in the fall, brand them and turn them loose for the winter to be gathered in the spring round-up and then taken to the home range for calf branding and beef shipment in the fall. Now this mass of cattle with many different brands and ownership was fleeing, freezing, starving too spent for any sound, but that pitiful lowing moan as they struggled on and on.

From then on snow lay on the ground almost constantly for many weeks. A few days of warm thawing weather always followed by more snowfall, the cattle perished by the thousands. We had sheds for the sheep but in the extremely cold nights they would pile up in the corners and smother the ones underneath, so that each morning there would be a number of dead ones. We girls sometimes went out in the afternoon to help skin them but never proved expert at it and our help counted for little.

Much of the time there was no grazing except sage and the tallest and least nourishing grasses. In order to save the flock from starvation, Father hauled shelled corn from Dodge City for them that winter. The cold and exposure he suffered in doing this permanently injured his health.

Before spring a new industry developed, that of skinning the dead cattle that lay everywhere over all the country. At the ranches when the fall work was finished, many of the cowboys were laid off for the winter. These men were welcome to stay on and be fed, but they spent the time drifting from ranch to ranch, visiting, happy and carefree. The most of these, after obtaining permission from the cattle owners, now turned to skinning and they made good money at it too. Tons and tons of hides were hauled to Dodge City and other points. Later the bone pickers reaped another harvest, but these were not buffalo bones.

The blizzard took a toll in human lives also. Many of the settlers lived in little board shacks that had almost no protection from such a storm. Most of them had scarcely any fuel except cow chips. Some of these people burned every piece of furniture they possessed. Some tried to reach their neighbor's homes and these all perished in the intense cold and blinding snow. The dugouts were veritable "places of refuge" at that time.

The second blizzard was the following year, and like the first, it lasted for a day and night. Again it was a stinging whirling mass of fine snow, carried by a high wind and terrible intense cold.

We had raised some feed that year and had a long stack of it on the west side of the corral. The shed extended across the entire north side which made quite a shelter from the storm. The feed had been stacked up close for just such an emergency, for the previous hard winter had proved a lesson to all to make what preparation they could. Only brother Will was at home to care for the sheep as Father, being on the Board of County Commissioners of Meade County, had been to town for several days at one of their meetings.

The snow started falling at daylight. About mid-forenoon Will decided he had better try to get some feed over the fence to the sheep and was gone so long that we in the house got uneasy. I put on overshoes, overalls, coat, and warm hood and tied a cloth over my face so only my eyes were left out. I sallied forth to see what was the matter but I couldn't face that storm even as bundled up as I was. It simply snatched one's breath away and the stinging snow filled the eyes so there was no sight. So turning my back to it I worked by way backward until I reached the corral fence and was much relieved to find brother Will inside working busily.

Part 1 • Chapter 9

When I told him that we had been worried about him, he said, "Gee, don't worry about me, I'm all right as long as I stay in here and keep working I won't freeze. But I've got to keep these blamed sheep moving or they will pile up and smother by the dozens." So I made a run for the house, going with the storm, to tell Mother and the three younger children not to worry about us. Once again I backed my way step by step to the corral to help keep the sheep moving.

All day we worked, carrying arm loads of feed to the sheep for a pitchfork was useless in that tearing wind. And we kept the sheep moving around and around the corral for exercise, poking them out of the corners of the shed where even then they were piling up in pyramids smothering those underneath.

At times we were forced by sheer weariness to crouch down for a few minutes and rest in the shelter of the feed stack but it was only for a few minutes for the cold was so intense we would soon feel chilled to the bone and so were forced to keep on the move. Toward evening I became quite exhausted and had to go to the house, but Will stayed with them until after dark when he had to leave them to their fate. I do not remember how many of them were dead the next morning but it was plenty.

Afterwards when we told people we had been out for most of the day, they could scarcely believe it. But, of course, the shed and feed stack made quite a windbreak else it would have not been possible. Father got home the following afternoon and was thankful to find us all safe.

He had been only two miles from home during the blizzard, having come out from town with a couple of bachelor neighbors the night before the storm. He had intended to walk home in the morning, but when morning came that was impossible. The boys he stayed with had their team in a dugout stable a very short distance from the dugout they lived in. One of them had attempted to get to the stables to feed the horses the afternoon of the storm but had to give it up and returned to the house with a frozen cheek and ear.

All old timers who experienced these blizzards will agree with me that these were more severe than any we have had since then. But from that time on everyone made better preparations for the winter. The mild weather of the first few winters had been misleading and most people believed there would never be any severe storms. The old buffalo hunters however might have had a different story to tell had they been questioned on the matter.

Schmoker house east of the Stone Schoolhouse. From left: Art Schmoker, Ella Schmoker Orr, Tom Johnston (Elisabeth's husband,) Will Schmoker, Elizabeth Schmoker Johnston, Children: Anna and Mildred Johnston, Mrs. Chris Schmoker, and Elmer Schmoker. Photo by F.M. Steele.

Chapter X

It was inevitable that the early settlers should have some Indian scares. The Northern Cheyenne's had passed through the country in 1878 and committed a number of depredations. Therefore occasionally rumors reached us that the Indians south of us were again getting restless and making ready for another break in further attempt to achieve some of their lost liberty.

One evening, the first summer we were on the sheep ranch, three cowboys rode up near the house and held earnest conversation with Father. They urged him to take his family to a place of safety at once as a band of redskins were on the warpath and headed this way. Father questioned them closely as to the source of their information and then quietly announced that he was willing to take the risk of staying right there. It was his opinion that this was just another wild rumor without any foundation in fact.

"All right," said their spokesman, "that is for you to decide. But we are just going around to all the ranches where there are families and warning them." Father thanked them and they rode off. It was suppertime but we had lost all appetite for the meal. Father made light of the whole affair, saying that it was probably only a ruse of some of the cowboys to scare sheep men out of the country. But when bedtime came he brought his axe inside the house as our only weapon of defense, for if we owned a gun there was probably no ammunition for it. He then barricaded the outside door of our log kitchen.

I do not know if my parents slept at all that night, but I am sure we older children were quite wakeful. When I did doze off it was always to have some horrid dream of fleeing before a hostile tribe and looking back to see the red flames of fire destroying our home.

When a peaceful morning dawned, bright and clear, we felt much safer and when a second night passed without molestation, we felt perfectly secure. But a few years later when the country was well filled with eager new settlers, there was an Indian scare of which we were blissfully unaware until some time after it was all over.

We heard it first from a cowboy friend who came visiting to our house and with twinkling eyes and a look akin to that of the cat that caught the canary. He asked us if we had heard of the big Indian scare the settlers were having. Of course, we had not and on being pressed for details, he gave us no real information. But he let fall a few remarks that gave us to understand some mischief had been a foot. A hoax that even the perpetrators had not foreseen the wide spread terror it would inspire and that it was just as well that the truth was not known. So we asked no further questions, however, in time we came to know that the solemn faced Kid Hobbs and a few kindred spirits had in a spirit of devilment wrapped themselves in red blankets and rode Indian file along some of the ridges and hilltops adjacent to homes of a number of new settlers. Only that and nothing more but that was sufficient to start one of the wildest and most terrifying Indian scares that could be imagined. These new settlers fled in mortal terror, spreading the news as they went. "The Indians, they were coming and we have seen them, flee, flee for you lives!" was the word they spread, stopping only long enough to shout the warning. The word was carried on and on and with each repetition the story grew and grew until it reached far and wide.

With frantic haste teams were hitched to wagons, some took along their few scant belongings, women and children were piled in and then most of them leaving with teams at full gallop. They drove on and on, the very wind seemed to carry fear and transmit it from one fleeing family to another. Several hundred gathered at one of the little new towns and prepared to make a stand against the dreaded foe. Others, many of them, pressed on to Dodge City.

After a day and night of fear and excitement, calm and reason began to return. One by one or in groups, these families began to trail back to their abandoned homesteads, but there were some who never returned. They just kept on going back east to civilization where such things as Indian uprisings were unknown.

When the truth finally leaked out, very little ever was said about it, for the men were rather shamefaced, realizing they had made their own exaggerated fear out of a mere nothing. The season of 1886 was another encouragingly wet one. Some crops were raised, every quarter section that was considered fit for farming was filed on.

New towns had sprung up here and there, as if by magic. Things were moving, especially in the real estate offices. In the rapidly growing town of Meade Center a school was arranged for and by late September of 1886 it started and I was going to school there.

Part 1 • Chapter 10

Our friends, the Boyer family, made this possible for me. Mr. Boyer had built a neat little three roomed house in town, especially for the use of his children during the school term and my younger sister and I were invited to share this home with the three Boyer children. Meda and I were about the same age and chums so it was arranged. My sister however was only with us for a couple of months and then had to return home because of a severe attack of 'flu'.

There was no school building in the town, there had been neither time nor money for it so the Christian Church was used for school purposes that winter. The main auditorium was divided through the center by heavy curtains that hung from the ceiling to the floor. In these two rooms were crowded the advanced and intermediate students, while in a side room efficient young ladies taught the primary grades.

I will never forget the confusion and difficulties of the first few days. Pupils, from nearly every state in the east, came with the textbooks they had brought with them from their former homes and some with none at all and we tried to have classes.

However, in time adjustments were made and we settled down for real work. One had to be able to concentrate in order to keep one's mind on their own lessons and recitations in that school. With the noise and recitations going on both sides of the curtain, it was as if there had been nothing between the two rooms. These harassed teachers certainly had a hard time, I know, for once when the intermediate teacher was ill I was chosen to substitute for her. When she finally was able to return to her duties, I felt like a complete wreck.

We found that by getting our lessons at home in the evenings we could get along very well so our school work was very satisfactory. Weekends the Boyer children went home and I sometimes went with them. But most often I went out to Uncle John's and that was fun for there was always plenty going on there.

These were the days when the Literary Societies were in their most popular times. Friday evenings there was always Literary, either at the Belle Meade, or Lake View schoolhouses and we all piled in the big wagon and went. Mr. John Innes was then one of Meade County's outstanding teachers, he and another young man, Mr. Christoffels, were the leading spirits in the Literaries. Mr. Innes excelled in the debates that were always a main part of the program, and Mr. Christoffels always gave readings and sang for he had a good baritone voice. One of the things that made these Literaries a real success was the willingness with which every one responded when called upon to take a part. Some of the attempts were of course quite crude, but that made no difference in the applause, they did what they could and to the best of their ability. Therefore, they were given credit for the effort.

We had a lot of fun at those Friday night Literary meetings. They served to bring together all the people of the communities on a friendly footing and the contacts did much to promote neighborly goodwill among the homesteaders.

The following summer, 1887, the Stone School House was built near our home by neighborhood subscription and donated labor. It was there that I taught the first term of school; a three-month's term at $20.00 per month. My pupils were my two younger brothers and sister and a number of children of our German neighbors.

Twenty-two pupils were enrolled and I was quite busy. But teaching was not a new experience for me for I had been teaching the younger children at home for several years as well as some of the children of near neighbors who came in to share lessons with my brothers and sisters.

It was my plan to keep on teaching but that winter Mother's health failed and for several years I was both nurse and housekeeper at home. When she was better I taught one more term in the little Stone School House, a longer term and the pay was $25.00 per month. At that time only a few districts paid as high as $30.00 a month.

From 1884 to 1887 were the 'boom' years. The cattlemen were being driven from the country. Several moved to New Mexico, among them the A.H. Bar outfit whose ranch was near ours. Mr. Johnston, the foreman, helped move the cattle to New Mexico. He then returned to this country and in the spring of 1887 he and my sister (Elizabeth) were married and made their home in 'No Man's Land'. Here in No Man's Land a number of ranchers still held out, as this land was not open for settlement.

When it became known that no laws were enforceable and no officer could arrest a person in No Man's Land, there was a time when law-breakers flocked to that haven. Then quite often persons stopped at our place making inquiry as to how far it was to the line on No Man's Land. Most of them seemed to be in a great hurry to get there.

One couple in particular I remember, for they stopped at our house for dinner. They seemed so nervous and ill at ease that we put them down mentally as elopers. Once we entertained a real bridal party at dinner. This was John Peckham, a cowboy from No Man's Land, and his bride, who had been Miss Jessie Slover, the daughter of one of the new settlers on the 'South Flats' between the Cimarron and Beaver Rivers. They were accompanied by a best man and a bridesmaid, but their names I do not recall.

Part 1 • Chapter 10

The couple had been married at Meade, Kansas the evening before. They were now on their way back to their home in No Man's Land, a long day's drive in the horse and buggy days. Our ranch was about the halfway point, and it mattered not at all that we had no inkling of their coming until they drove in at almost high noon. Nor did it matter that we had never before set eyes on any of this party. Their names were familiar to us through the talk of mutual friends and in the same way they knew of us. So when introductions were over we gave them whole hearted welcome and the best the larder afforded on such short notice. The bride looked very 'bride-ish' in pure white from head to toe with one of the large picture hats of that period and the bridesmaid was in equally appropriate attire. I am free to confess no small detail of their costumes was overlooked by the feminine part of our household and was to me a diverting change from the ordinary routine.

Many strange and unusual characters drifted about in those days, some were pathetic, as for instance, the Van T brothers, Germans with a limited use of English. They lived considerable distance from us but when they found that Father could speak the German language well they came often to our home, always spending the entire day in conversation in their own language. Their talk was all of Germany; to them it was the most wonderful and the most perfect country on earth. No matter what came up during these talks, everything was compared with things as they were in the old country and always to its credit.

We easily gained from all this talk that they were terribly homesick. We often remarked after one of these visits, "Why don't they go back to Germany and stay there since everything there is so much better and more desirable than here?" And then after a time we got the answer to this question.

Their's was a family of some consequences in the old country. The Mother was a widow with some means and when the time came near for her boys to enter the German army for the usual term of years that was required of all young men of that nation at that time, neither of the boys had been willing to accede to this compulsory training, nor was their Mother in sympathy with the war policy. So the boys slipped away to America without permission from the reigning powers, and then found that they could not return as they had expected after their age of service had passed, without suffering a severe penalty. So longing hopelessly for their homeland they stayed on and on, living miserably on a small remittance sent them regularly by their Mother. Without hope or ambition or interest in this 'Land of the Free', they existed aimlessly.

The older of the two died a few years later somewhere in No Man's Land with his longing for the homeland unappeased and the other drifted out of our knowledge.

It seems almost a marvel to me now to remember how some of those first settlers lived. How little variety there was in the food they existed on, especially some of the bachelors we knew.

One of these was a queer old fellow who lived near us for several years, living alone with his dogs, a few beans and a queer mismatched team. He managed to raise a little corn during the good years while the rainfall was sufficient and on this they all lived. He always kept a pot full of boiled field corn on the stove. Whenever any of the cowboys from the nearby ranch or my brothers stopped in to see him, he always urged them to stay and eat with him. Setting out the boiled corn and some muddy coffee he would say, "I am a poor man but eat, eat."

He and the dogs seemed to thrive on this fare, but he was very fond of buttermilk and twice a week came over to our house for a pail of it. Mother was glad to give him all she had to spare. When handed the filled pail, he always drank about half of the contents before leaving the place and then going by the spring would refill it by adding that much water. He claimed the diluted buttermilk was just as good as the whole buttermilk. Poor old fellow, we always felt sorry for him and yet that did not deter us from having some fun at his expense on one occasion.

He had a fine melon patch down on the creek bottom, some distance from his soddy. When they began to ripen he kept complaining that some one was getting into his melon patch every night and destroying the best melons. As a matter of fact no person had touched his melons, but the coyotes were the offenders. The old man could not believe that and announced one day that he had a gun ready for the fellow who was 'monkeying' with his melons.

The talk of that gunplay gave us a brilliant idea. Three of us young folk decided he needed some excitement and so did we. So that night, soon after dark, we saddled our horses and sallied forth. Just before reaching his place, our companion struck off down to the melon patch. My sister and I continuing on down the road that led close by the old sod house. A light gleamed within; nearby the old team was tethered out on picket ropes. We dashed past the house at a mad gallop, the cur dogs yelping at our heels, the old team excited at the hubbub snorted wildly and ran on their ropes. It was then we got thrills aplenty for the old man came running out, gun in hand, yelling loudly for us to stop. Instead we left the road and raced off through the tall sagebrush and to the hills beyond, the old man's threats and shouts following us for

Part 1 • Chapter 10

some distance. We reached home by a round about way, where we found our companion who had reached home a half-hour before ahead of us although encumbered by a huge melon. I wonder why a stolen melon is so sweet? That one, I am sure, was the best I ever tasted. The next day the old man had an extra tale of woe for casual callers. He said some fellows had tried to raid his melon patch and he came near "getting" them, but they out ran him and got away.

Cowgirls at the foot of "Lover's Leap." Taken about 1888, from left, Carrie Schmoker Anshutz, Nell Graves, Ella Schmoker Orr.

Chapter XI

During the inflow of settlers from 1884 to 1887, the seasons were quite encouraging, but these were followed by dry seasons with hot winds, which burned up the crops. Many of the homesteaders finding living conditions too hard to cope with, began to make plans for leaving the country. In the fast growing little boomtowns of those days, many loan companies, using eastern capital now, did a thriving business. For hundreds of homesteaders found it possible through these loan companies to "commute" on their homesteads, that is, pay for the land at the rate of $1.25 per acre. This money, with something additional being furnished by the loan company, and a mortgage on the place given them as security on the loan, the amount of course varied according to the lay of the land, but $500.00 to $600.00 was about the average. Many of the discouraged settlers made final proof, secured their loan, gave a mortgage and left the country the next day.

The loan companies took these mortgages east and sold them. A great many of them were sold to farmers, mill workers, servant girls and widows with a few hundred dollars to invest.

This land left to lie idle produced not one penny of income. Taxes became delinquent, interest on the mortgages went unpaid, and the little boomtowns began to shrink. Only the people with livestock to depend on were able to stay with the country.

The boys in our family were tired of sheep herding, so the flocks were sold and cattle purchased instead. The small ranchers, located on streams, had taken the place of the big cattlemen. They kept track of their herds by daily riding, the cattle roamed over the deserted homesteads of the high flat country and the hilly country back from the creeks. Most of the permanent settlers added to their holdings, some of the land was now offered at such a low price.

In 1892, I married M. W. Anshutz who for a number of years had been foreman of the Taintor Ranch in No Man's Land. This land was now open for settlement and our new home adjoined that of my sister and brother-in-law, the Johnston's, on the Cimarron River.

When my husband to be, consulted me as to what manner of house to build for or home, whether of sod, stone or lumber. I answered without hesitation, "Anything, so long as it is not a dugout and has a roof that does not leak."

So ours was a white painted five-room house (frame) moved out from Englewood. In those days there were many vacant houses in the towns and country, people bought them and had them moved to their ranches. At my Father's place, at about this time, two houses were moved out from Meade and the big, old half dugout and the log kitchen in front of it were abandoned and the family moved into the roomier and better looking frame house. But the old house holds a very tender place in my memory, even though the roof did leak on occasions. It has been said that no home has served its time until under its roof there has been a marriage, a birth and a death. My marriage took place there in the old home. A niece of mine was born there and my father died, just a short time before the family moved into the new house.

I was very proud of my frame house with its smooth white plastered walls, its painted woodwork and fine new furnishings but as to the roof the joke was on me after all. No, it did not leak when it rained but we had a snowstorm that year, the hard powdering kind, driven by a high wind, it found its way into every crack and crevice. A lot of it got into the loft, over our small kitchen, for the shingles had shrunk from long drought and a few perhaps had blown off. It was very cold the day following the storm and we had a rousing fire in the cook stove, the kitchen was warm as toast.

My husband, a visitor and I were sitting near the stove talking about the coziness of the room. Presently the two men moved to the other end of the room, when BAM without warning, almost the entire plastered ceiling fell and all of it that could, came down on my head. What raised my ire was that the two men laughed. They had just stepped out of the danger zone before the crash came, and I suppose I was a sight with my face and hair plastered with lime dust, but they said it was the look on my face that sent them off into peal after peal of laughter. The heat from the stove had melted the snow in the attic and we had not noticed the damp seeping through the plastering so the crash came so unaware. For a minute or so it was hard to realize what had caused the catastrophe.

We had many visitors. In those days folks just rode or drove up and stopped overnight or for the noonday meal without questions. The cook never knew how many might have to be served before mealtime was over. That did not trouble me, for I always had been used to this, as it was the custom of the West.

Part 1 • Chapter 11

Only at very rare intervals was any one seen traveling through the country on foot, but one morning a middle aged foreign looking man stopped and asked to be taken across the river. While my husband went to get up a horse for him I talked with the man for he had an earnest intelligent look. He said he was trying to get to some friends who lived in southeastern Oklahoma. He spoke broken English and I asked, "Are you German?" "No, Madam," was his answer, "I am Russian."

"Tell me," I said, "something of the conditions in Russia. Are they as bad as we sometimes read of in the newspapers?" I suspected he was a refugee. "Russia," he replied slowly, "is a wonderful country, the best country on earth. But in order to be able to live in it, a man must have no politics and no religion." And the terse but enlightening statement quite satisfied my curiosity as to our unusual traveler.

One morning a group of four or five horsemen passed down the Healy Trail, a short distance to the west of our house, following them like a dog was a loose horse. I wondered at this. Cowboys always drove their loose horses ahead of them.

When my husband and my young brother rode in a short time later, I asked, "Did you see the outfit that rode by here awhile ago. Who were they?" "That is what we would like to know," my husband replied, "but we did not think it healthy to find out for when we started to ride toward them they went through the gate across the river then motioned us to go around and waited with gun in hand until we did so."

We had not long to wait for the solution of this mystery for a sheriff's posse was hot on their trail. These were a band of bank robbers, they had held up and robbed a mail train and killed the messenger on the Santa Fe near Cimarron, Kansas. They were making their way back to the wild shelter of the Wichita Mountains, from which they had come, for they were seen passing through a few days before on their way to the planned robbery.

They stopped at the Taintor Ranch for dinner that day, one of the men stayed at the barn with the horses while the others ate. About two hours before the robbers came to the ranch, a deputy sheriff had been there looking for them.

"Better stay awhile," Mr. Taintor remarked jokingly, "they may come in." But the deputy said he had better be getting on. So, of course, when they really did come, an hour or two later, the ranch outfit was quite well aware of whom they were entertaining. But these men were desperate and were heavily armed, so they were given food and allowed to go on.

On the Kiowa, they were overtaken by Frank Healy, Sheriff of Beaver County, Oklahoma, who engaged them in a running fight. A horse was shot from under one of the men, probably the extra horse was for just such an emergency, but they made their escape and were never apprehended.

Some time later we had an experience with a lone lawbreaker. It was a quiet Sunday evening in late summer, we were sitting on the front porch enjoying the cool breeze and the brightness of the full moon when a stranger rode up and asked the way to Englewood. My husband walked out to put him on the right road.

He said he was from Beaver City and that he had received a telegram that afternoon telling him of the serious illness of his mother somewhere in the East. He was trying to reach Englewood in time to catch the early morning train to the East. With the quick sympathy of true westerner for one in trouble, my husband, although the man was a total stranger, offered him a fresh horse for the remainder of the journey. This offer he declined and it was his undoing, for about five miles farther on, at the E.P. DesMarias Ranch, the horse he was riding gave out and he was forced to stop for the night.

About two o'clock that night, we were awakened by the under sheriff from Beaver, who with a couple of deputies, was on the trail of the young outlaw, Fleetwood. He had been in the Beaver jail for some time on a charge of horse stealing. On this Sunday afternoon an exciting game of baseball had been in progress when this young fellow, in the absence of the "law" due to their participation in the ball game, managed to make his escape and stealing a saddled horse tied to a hitching post made his break for liberty.

When the ball game was over and the prisoner's absence was noted the under sheriff, without taking time to change his ball suit to other clothing, was hot on the trail of his charge. Had Fleetwood not pressed his mount so hard in the beginning of his flight or had he taken advantage of the kindly offer of a fresh horse, he might have made a get-a-way but as it was they found him in bed in the bunkhouse at the DesMarias Ranch very early that morning.

Our own breakfast was over and our men were starting to work in the hay field when the under sheriff with his deputies and the prisoner returned and stopped at our house for breakfast. The under sheriff was rather testy of temper and looking undignified, if not ridiculous, in his abbreviated ball suit of bright red with white trimmings. He held his gun on the prisoner while he was forced to do the work of caring for and feeding the horses at the barn while I prepared breakfast. During the meal the prisoner, far from being downcast at his recapture, indulged in various wise cracks at the expense of the officials on the law, which did not serve to diminish their ill temper. Later on however this gay young outlaw made another jail break and that time he was successful in not being recaptured.

Life in those early years never became commonplace or dull. Almost every day brought new experiences and new faces. The Kansas line was just a mile north of our home, along this line practically all of the homesteads had been mortgaged and abandoned. After a number of years of unpaid interest on these mortgages, investigators were sent out from the East to inquire into conditions and to find the reasons for these delinquencies. It was probably the second summer after our marriage that one of these mortgage investigators was a chance visitor in our home.

He had been sent to Meade, Kansas from one of the New England states and this was his first trip to the West. Securing a livery rig and a driver at Meade, he was out making his round of investigation. Several of the places on his list were along the Kansas line, just north of us. The driver, who was Jack Roberts, found it convenient to drop in at our place for dinner. Jack brought his man in and introduced him with the air of one introducing a huge joke. Jack himself seemed filled with suppressed laughter.

When we gathered about the dinner table and the men began to talk, I soon realized why Jack was in such a state of mirth. He had been listening to this man's talk all the morning. Jack, like ourselves, had grown up in the West and like all Westerners took things as they came as a matter of course. But this man, perhaps sixty years of age, had never been outside of his New England environment and he was shocked, horrified and dismayed by this Western world.

Never in his wildest dreams had he ever visioned anything like it. I will admit he was seeing it at its worst for it was a very dry midsummer and a hot wind blowing. But what caused my husband and Jack to exchange understanding smiles was that over and over again he deplored the "lack of society." From the bottom of his heart, the poor man pitied us all because of our lack of society. "Why," he said, "we passed a school house away back yonder some where between here and that little town we left this morning and six or eight poor little children stood out there in the hot sun and watched us drive by, not a tree was in sight, nothing to play with, poor, poor little creatures." And he shook his head as if too much overcome for further words.

The two men, highly amused, led him on and on, but as I listened and looked him over I became somewhat indignant. Why should this man criticize our country and pity us who lived in it? The Western men, as a whole, were bronzed and bright eyed and healthy looking. But this stranger was the exact opposite. He had lusterless eyes and his skin was sallow and unhealthy looking. He was narrow chested and stooped and he had an asthmatic wheeze and a nasal twang. So I arose to the defense of my country, I told him we liked living here. The school children that he pitied so much were most likely happy as larks.

He looked at me in amazement. "But don't you miss society?" he asked. "No," I replied, "for I am perfectly happy and content without society. Why should I not be, for here I have lived for many years. I am surrounded by friends and my family and relatives are here. We have our health, comfortable homes and a fair living. What more could any one want?" "Yes," he agreed grudgingly, "that was true enough, but the lack of society..." and he was off again on his favorite subject.

It was at this point that Jack hastily excused himself and went outside where he could laugh unrestrainedly. When my husband joined him a few minutes later he said, "That poor old nut has been going on like that all morning." My guest and I continued our argument for some time but it was plain to me that here was a person who could never be convinced that the West held anything worthwhile, could never look upon it with compassion or understanding.

When he came to take his leave, he extended his hand and said, "Mrs. Anshutz, I am very glad I met you. I shall never forget you for I shall always be glad to remember that in this awful, lonely, desolate country, there is 'one' woman who is happy and content." And, he emphasized the one as if he were quite sure that there was but one female with such a strange choice for the place to call home.

After thirty years of life in the West, my sister, Elizabeth, and I returned to the home of our early childhood in Iowa for a short visit. It was October and the wooded hills along old Pine Creek were glorious in their colorings of crimson, green and gold.

We took long walks and drives to all the old familiar haunts, the old home, the church, and the schoolhouse. Went hickory nutting in the woods and wandered once more amid the wild beauty of Wild Cat Den, where great pine trees grew out of seemingly solid rock cliffs.

We never tired of watching the steamboats and motorboats on the broad Mississippi and we enjoyed to the utmost meeting our friends, schoolmates and the relatives we had not seen in so many years. But when it was time to return to our homes my sister voiced the sentiments of us both when she said, "Well, it has been grand to see all the old places once more. The hills and woods are beautiful but I could never again live here and be happy and content." The West had put its brand on us. We belonged there.

The years have brought about vast changes, there are no more frontiers as we lived in during the early times of our life in the West. The telephone, motorcar and radio have drawn even the most isolated places closer.

This was brought to my mind while vacationing in New Mexico and Colorado a few years ago. We left Taos one afternoon and traveling north over a road that was not well traveled. We passed through a very sparsely settled country for many, many miles. "This," I told my daughter, "reminds me somewhat of the early days in our country except that here the vegetation is sparse while in ours it was rankly luxuriant." But there was the same look of loneliness, the houses were miles apart, most of them were shacks, the inhabitants Mexicans and Indians. A few sheep and goats grazed about, burros and small hardy ponies seemed the usual mode of conveyance. The whole scenery reminded me vaguely of the past. It was like and yet unlike my own early surroundings, but as we stopped by the roadside to photograph a lovely blooming cactus, a motorbus passed us. My glance swept past the driver in his smart uniform to the passengers. A group of Indians, the central figure of which was a woman past middle age with a look almost of death on her face, with closed eyes her head rested on the breast of a stalwart young man, evidently her son, and his strong arms supported her. As they passed he turned toward us, such a look of hopeless sadness, as I did not believe the stolid Indian countenance capable of expressing. It was a glimpse of desert tragedy, but I was impressed by the evident fact that even in this isolated spot these people now had recourse to the swift motorbus in time of sickness and trouble, for obviously the sick woman was being taken to some Colorado town for hospital and medical care. Modern efficiency available in the desert.

It was because of the changes the years have brought that I wished to leave some record of the times now long in the past, for their like will never be seen again or never again can such conditions exist. The "Drama of Life" in those years contained much of comedy and lighthearted laughter, but there was also tragedy, mystery, romance, and enveloping it all there was a charm, a fascination that none will deny. I had hoped in these pages to give some glimpse of this, perhaps I have failed, the charm of those early years too elusive for my pen, the fascination of them something I cannot explain.

But my husband's story follows, his horizons were wider than mine, perhaps the picture may be made plainer. I shall write the story in the first person, as he has told it to me.

Carrie Anshutz with daughter, Esther, shelling peas by their home.

Carrie Schmoker Anshutz standing in front of the Stone School in southern Meade County.

Part 1 • Chapter 11

Will Schmoker's daughter, Fannie Schmoker Judy, about 1907.

M.W. (Doc) Anshutz

Part Two

M. W. (Doc) Anshutz's Life Story

As told to Carrie Schmoker Anshutz

by Doc Anshutz in the year of 1936

Chapter I

My boyhood was spent in Bellaire, Ohio, the town where I was born on July 21, 1861. My parents were among the pioneer families. Father was born in Germany and as a young man was closely associated with Carl Shars, noted German American political leader, and with him was engaged in the revolutionary movement of 1848-1849, in Germany, as a result the leaders were forced to leave Germany. Coming to America, the Anshutz family settled at Bellaire, Ohio. Father was an iron molder and engaged in that business, while some of his brothers became river men, running a line of steamboats that plied between Cincinnati, Ohio, and Pittsburgh, Pennsylvania.

Mother's people, the Nicolls, were Massachusetts's Yankees. Her father served in the War of 1812, enlisting when but twelve years of age. He was in the battle in which Tekamah, the Great Indian Chief was killed. Many stirring tales of that war were told us by grandfather, who to the day of his death carried a bullet in his hip as a constant reminder of his wartime service.

Coming to Ohio they were among the first settlers in that part of the state and they engaged in the lumber business. Grandfather owned and operated saw mills and planing mills at Bellaire. My brothers and I were taught to do everything about the business of the mills, for my father died when I was about two years old, leaving mother with a family of young children. We then went to live at Grandfather's and grew up there in a large comfortable, old-fashioned house of sixteen or more rooms.

My boyhood was a pleasant one, there was the usual schooling, of course. But I much preferred the work about the mills, then, too, Grandmother kept a large greenhouse, purely for her own pleasure.

She saw to it that I served an apprenticeship in it, for children in those days, were taught all manner of work. Although there was time for play too, and during the summer months we boys spent much of our play time about the river, so that even as a very small boy I could swim like a fish. In winter the sport was ice-skating. So the time passed with school and work and play until I was nearing sixteen years of age. Then quite by accident I had my opportunity to come to the West that has ever since been my home.

I had an uncle, my mother's brother, who had come out to Spearville, Kansas, about fifteen miles northeast of Dodge City. He settled on a homestead there, and after making some improvements in the way of living quarters, he sent for his wife and small children to join him.

She prepared for the journey and just the evening before her departure Grandfather decided she needed someone to go with her to help with the children and the luggage. I was asked to make the trip out with her and in a short time return home. Always eager for new scenes and experiences this suited me immensely.

It was in April 1877, that we came to Spearville, then just a mere station on the Santa Fe. After a short stay with my uncle's family I felt that I could not go back home without seeing more of this great new prairie country. And so quite naturally I soon found my way to that "Mecca" of the West, Dodge City, where I made the acquaintance of "Prairie Dog Dave", an old buffalo hunter. He invited me to go out with him on a buffalo hunt.

I seized the opportunity eagerly, what boy would not? For, besides the chance for sport that it offered, it meant camping out and seeing a lot of the new wild country that lay in such great stretches on every hand, all so unlike the settled country of Ohio and West Virginia with which I was so familiar.

So with team, wagon, hunting and camping equipment we traveled south from Dodge City as far as the mouth of Crooked Creek, where it empties into the Cimarron. There we turned west, going in that general direction until we found the Jones and Plummer trail. Sighting only small and scattered bands of buffalo, for by then the great herds had been thinned out by the ruthless slaughters, and the remnant of those once "thundering herds" had been driven farther and farther away.

Disappointed in the lack of game, my companion decided to return to Dodge City. So as a buffalo hunt, the trip proved fruitless, but I had here my first look over what was later to become very familiar ground to me. And I wish that I will be able to furnish my readers with a description so accurate as to bring them a real mental picture of the

Part 2 • Chapter 1

country that unfolded to my view as we plodded on our slow way from day to day.

A land of opportunity lay here untouched by the hand of man. Nature had spread here a rich store, for the buffalo grass on the uplands lay a heavy carpet of rank growth. While on the low lands, the grass everywhere was knee high and along the streams and the meadows, back of them, the ranker meadow grass waved in many places as high up as a horse.

The main streams were not wooded, but back out of the line of flood waters were scattered many stately cottonwoods. The small spring-fed tributaries of the main streams and many of the deep draws and canyons were more heavily wooded, which offered shelter from winter storms and grateful shade in the summer to the denizens of the plains country.

Thousands of wild plum thickets grew along the streams and draws. In early April they were white with bloom and their fragrance was carried afar in the clear sweet breeze.

In many sandy places grew the sand hill grapes, these grew in short straight bunches with many trailing tendrils. The bloom of the grape came on later than that of the plum and the scent was a faint sweet elusive one.

By early July the sand plums began to ripen and from then until mid August there was a rank abundance of this delicious fruit. After it came the ripening of the purple seedy, little wild grapes and these were the only fruits the country offered. No doubt, the Indians had for centuries enjoyed these fruits, prepared for them by old dame nature, just as did the white people who came later to make their homes here. For as with the Indians, these were the only fruits these first settlers had.

At this time, 1877, there were no ranches or cattle in all the country we traversed. For as yet, the whole vast expanse for many miles to the south and west lay open, un-grazed except by a mere fraction of the once vast herds of buffalo.

Here and there scattered bands of antelope, which dotted the open plains. In the rough hills and wooded draws there was an abundance of deer. By night the gray wolf skulked and coyotes sent out their long lonely wails.

Over the high prairies were numerous prairie dog towns, their cheerful and friendly little inhabitants always ready with a saucy challenge to the passerby, whether animal or human kind. And living in their dog towns was always to be found the deadly rattlesnakes. The odd harmless little owls lived and bred and raised their young in some of the subterranean labyrinths of the prairie dog towns. In the spring and early summer their call of "cuckoo, cuckoo" was clear and musical on the evening air and it held a note of wistful loneliness that fit well with the immensity of the prairie reaches over which it echoed.

The broken country back from the streams was filled with allure, for they offered continual change of scene. What might be the view from yon distant hilltop would be hidden in the dark depths of the deep draws. The long stretches of flat prairie were made enchanting by the mirage that held such promise of tree bordered lakes of waters, which seemed to dance and shimmer in the sunlight. Then as we drew nearer to recede and take on some different form farther on.

This, my first experience of life in the open, was appealing, alluring and I had no desire to return to my boyhood home. This was the life for me. And so, soon after our return from this hunting trip, I found work with one of the herds of Texas cattle being held along the Arkansas River that summer. And I wintered on a tributary of the Saw Log, north of Dodge City, riding line and helping to hold a bunch of cattle there. So began my life as a cowboy just at the time when the days of the buffalo were over and the cattle industry, which was to become such an important one, was at its beginning in this part of the West.

The next two years of my life was spent mostly in the vicinity of Dodge City. I became acquainted with all the colorful figures that helped to make the history of that town so outstanding at that time. I also became quite familiar with many of the men and boys who brought up herds from the lower Texas country and with them served my apprenticeship as a cowhand.

From the first, my main interest was with the cattle business, and so I kept track of and was deeply interested in the establishment of these first widely scattered ranches that soon began to appear over this country.

The Dodge City that I came to in 1877, was young in years, rough and crude as to buildings, but it teemed with life and excitement. Money flowed freely for it was easy come and easy go.

A typical frontier town of the early west, it acquired a reputation and glamour that persisted for many a year. The population had been a shifting one of railroad builders, buffalo hunters, freighters, gamblers, adventures, a class of men not looking for wealth so much as for fun, excitement, adventurous living.

All of the business houses were on one street called Front Street, and were outfitting stores, hardware, drugs, saloons, gambling houses, dance halls, hotels, restaurants and boarding houses.

Part 2 • Chapter 1

The Ham Bell Livery Barns and Corrals were nearby, just south of the railroad tracks. In the residence part of town, farther up on the hillside, lived a number of families, the wives and children of businessmen of the town. They lived quiet and secluded lives and were seldom seen in the lower part of the town, where the scarlet women of the dance halls reigned supreme. Where at any time of night or day a "killing" might take place, for Boot Hill was then in the making. It was the place of burial for tough characters that met untimely and usually well-deserved death with their boots on and they were buried on Boot Hill. I was at one of the burials there.

Soon after coming to Dodge City, I was out south of town one afternoon with a Texas outfit who were holding a bunch of horses for sale. Some of the boys were preparing to go to town to spend the evening. I overheard the foreman admonishing one of his men, a burly fellow: "Now Jim," the foreman said, "be careful. Don't get in trouble over there in town." "Well," Jim answered belligerently, "I will be careful if they don't pick on me, but if they start picking on me there's going to be trouble." Someone must have made the mistake of "picking" on Jim, for that night he was shot and killed and the next morning I was among the spectators that witnessed his burial on "Boot Hill" for it was simply burial, and nothing more, there was no funeral rites, whatever.

Killings in those days were not, as some people were led to believe, of every day occurrence, but they were not unusual. I well remember the first one I saw. There was a large crowd in one of the principal saloons where much drinking and gambling was going on. A freighter and a gambler, who had had some previous trouble, met in the saloon and without parley opened up fire on each other. As usual in such a case, the crowd made a wild dash for the door and as it was not wide enough to accommodate all, a jam ensued in which I was knocked down and trampled by many feet. It was the freighter who was killed in this fight but he was not buried on Boot Hill.

Some of the happenings of those times, at these later days, seem fantastic and unreal, such a one is the following:

One day somewhere below Ft. Dodge, a man was shot dead and left lying where he fell by the roadside. Several persons saw him and came on to town and reported. Soon a dispute arose for several men reported it was a white man, but one man insisted it was a Negro. The argument waxed hot and in order to prove his point this man mounted his horse, rode back to the spot where the corpse lay, decapitated it and returned to town carrying the head of the victim, a Negro's head in a gunny sack.

But not all of the men in Dodge City at that time were tough characters, far from it, but they were an outstanding lot, the businessmen and main characters of those times. Some of these men were only casual acquaintances of mine, others I knew more intimately, such as Bob Wright and his sons, the Wright boys. Bob Wright was perhaps the leading businessman of the town. Wright and Beverly's, York and Drapers were the main outfitting stores. Later the Geo. Emerson Co. Store was established. "Chalk" Beeson, then a young man, later established a cattle ranch on Five Mile Creek, west of Englewood in Clark County. Mr. Beeson was also a musician and in later years organized the famous "Cowboy Band" that was known far and wide. Webster who was Mayor of the town, then there was Bat and Jim Masterson, Jim Campbell, Frank Brown, "Dutch" Henry, George Eddy, "Dog" Kelley, Andy Johnson, Billy Dixon, Jimmy Langston, and many others.

The Adobe Walls fight with the Indians was still fresh in the minds of all and I made the acquaintance of all the participants and heard the story first hand. Many of the mementoes of that fight were then on exhibition in one of the principal saloons of the city. I knew well Bill Tilghman, who was the City Marshal at one time. He was afterward made U. S. Marshall of Oklahoma.

Wyatt Earp and his brother were also well known. Dave Mathers, known as 'Mysterious Dave' was a descendant of Old Cotton Mather of Massachusetts. Mysterious Dave was known as a "killer" in those days.

I met Eddie Foy, the Actor, who was often in Dodge City in the early days and who was a great favorite in his line of entertainment.

Doctor McCarthy, the pioneer physician, whose practice embraced the entire Southwest; Old man Peacock, dance hall owner, a huge fat man, weighing well over two hundred pounds and his wife, who was even larger than he was; Alf Updegraff, popular man about town; Luke Short, Dr. Choteau, of the family of St. Louis Choteau's; Tom Nixon, City Marshall, who was later killed by Mysterious Dave.

I witnessed this killing, standing not fifty feet away. When the men met, I heard Dave ask Tom if he had his gun with him and then quick as thought, the fatal shot was fired.

P.G. Reynolds and his sons, George and Sidney, Mr. Reynolds was the stagecoach man, and his mail and stage lines were doing a thriving business of that time.

Geo Hoover, wholesale liquor man, Larry Didger, Walter Streter, Ham Bell, liveryman, who is still living in Dodge City at this writing (1936) and now well past eighty years of age.

Part 2 • Chapter 1

Ben Danills, who was also a City Marshal. He was later with Teddy Roosevelt's "Rough Riders" and distinguished himself as an expert marksman. These and many others I could name that belonged in those days.

Many of the old buffalo hunters too, were still there. In fact, it was the spring of 1878, two years after my coming to Dodge City, that I helped Andy Johnson, one of the hunters who had been in the Adobe Walls fight, to bale the last shipment of buffalo hides that was sent out from Dodge City.

I met and made the acquaintance of numbers of the buffalo hunters of the plains. They were in a class by themselves; they had a code of honor that was seldom violated. Their word was their bond. Grim and fearless in the face of danger, resourceful, friendly, honest, such were the buffalo hunters, as I knew them. Perhaps, in my boyish mind, I accorded them a measure of hero worship, their tales of adventure were so stirring, but their days were over now.

Much has been said and written about the ruthless slaughter, the stupendous waste that marked the extermination of the buffalo of the plains, but it was after all perhaps just part and parcel of the inevitable march of civilization to the West.

Chalk Beeson

Buffalo caught in "No Man's Land," Oklahoma in 1886 by Lee Howard for Buffalo Jones, Garden City, Kansas

Chapter II

Old Trails of the West

The old Trails of the West were an important part of its development. A study of them is most interesting. The best known and the one of the greatest importance here in the Southwest was, of course, the old Santa Fe Trail.

Dodge City was built on the line of this trail, but at the time of my coming, the railroad had replaced the trail and it was no longer used for its original purpose of transporting freight from the Missouri to Santa Fe and Taos in New Mexico.

Volumes have been written about the Santa Fe Trail, its history and the backgrounds of its development and yet the subject has not been exhausted. The pity of it is that so few of those early travelers of the Trail kept any record of their experiences and adventures.

Its beginnings lie way back in the dim past, when Mexico and New Mexico and, in fact, all of the little known territory lying West of the Missouri River was claimed by Spain.

The Spaniards were jealous of any interference, suspicious of the French and American traders and trappers who wondered by chance or otherwise into the vast stretch of wilderness that was then know as "The Great American Desert".

Some of the earliest explorers were met by Spanish forces and forced to turn back, though a few did manage to slip through. It would seem that this should have been easy to do when one considers the immensity of the open wilderness and that there were no means of communication between the Eastern part of the United States and those far off. Spanish possessions were to the Southwest, but it appears that the Spaniards were ever on the alert to intercept the Americans.

Coronado's journey in 1541, took him over at least a part of what was later the Trail. From then on through the 1600's and part of the 1700's history records only occasional white men who ventured across the wilderness.

The French were the first to endeavor to trade with the Spaniards, but with little success. So they traded with the Indians instead, but it seems they persisted in trying, for as late as 1795, the Government of New Mexico ordered the arrest of all French traders and confiscation of their goods.

Early in the 1800's a few American traders went, or were sent, to New Mexico with stocks of goods from Missouri points. Some reached their destination, others were robbed and some murdered by hostile Indians. The ones, who did reach Santa Fe, were seized by the Spanish Authorities, thrown in prison and their goods confiscated.

Zebulon Pike made his famous exploring trip in 1806. His published narrative in 1810, aroused great interest, not only for the adventure that lay in the West, but also for the profit to be derived if only it were possible to open up trade with New Mexico. So from time to time, efforts were made along this line with varying success.

But in 1821, there was a Revolution in Mexico and the power of Spain was broken, then the Mexican people welcomed the American traders and in the next two years several caravans made their way West along the Arkansas River and by the mountain route over Raton Pass and so on to Santa Fe.

All of these were on horseback with pack animals to carry their goods. In 1822 or '23, a few wagons made the trip and because of the difficulty of getting even horses over Raton Pass, these intrepid men then struck off from the Arkansas River and traveled across the waterless desert between that stream and the Cimarron, a good sixty or seventy miles at the best. Terrible hardships were encountered, they so nearly perished of thirst that they cut off their mules' ears and sucked the blood. Finally, more dead than alive they reached the Cimarron and so through heat and cold, thirst and hunger, Indian massacre and other hardships these men persisted against terrific odds for the rewards of the traders in that time were very great.

In 1823, the attention of the United States Government was called to this thriving and lucrative business of trading with the Mexican people. A bill was passed appropriating a sum of money to survey a trail from the Missouri River to Santa Fe, a distance of about eight hundred and fifty miles. Joseph C. Brown was appointed surveyor. Starting at Ft. Osage, on the Missouri, in two months the party had reached the Arkansas River, here a wait was made while permission was sought from the Mexican Government to continue on the other side of the river, this was granted and the following year the survey was completed.

Part 2 • Chapter 2

Joseph C. Brown did his work well, he made copious notes and drew an excellent map of the entire route, including the cut-off or shorter route that left the Arkansas somewhere near where Dodge City was later located and went on to the Cimarron. After 1830, this route was quite generally used for it saved several days time and dispensed with the crossing of the troublesome Raton Pass.

Brown's directions for following the trail were so clear that travelers could scarcely go wrong, but unfortunately, his records were not made public. Through some official red tape his excellent work was pigeonholed and forgotten and only came to light to be of historic value.

Had the frontiersmen had the benefit of his work undoubtedly lives might have been saved, but they continued their long journeys in the best way possible, some wandered off the track and got lost, for as yet, the trail was a mere 'trace'. But the year 1834 was a rainy one, mud was so deep the wagons made such deep ruts that thereafter these were plainly discernable and it was no longer necessary to get lost and die of thirst on the desert or "Jornada" between the Arkansas and the Cimarron. However, this cut-off was always the most dangerous part of the journey, for it was the one most often selected for Indian raids on the caravans and history records many bloody tragedies along this lonely way.

The Cimarron at the point where the Santa Fe travelers crossed it was often dry. But by scooping out holes in the sand, water seeped in and so sometimes death from thirst was averted only to be met later from ambushing Indians, shooting their deadly arrows at the unsuspecting wayfarers.

The greater part of the loads carried by the traders to Santa Fe was cotton goods of all kinds, silk shawls, looking glasses. Hardware commended fancy prices.

On the return trips they often carried furs, pelts, buffalo robes, silver and some gold bullion. Many mules also were brought from Mexico to Missouri.

Later, when the Mexican war came on, the Trail was in constant use by Troops, Infantry, Cavalry, Gun Carriages and a constant stream of supply trains.

In the gold rush to California, in 1849, many used the Santa Fe Trail for at least part of the way. But the traders, with their long trains of ox or mule drawn wagons, continued to make the long trips across the Plains, despite dangers and hardships, until the Trans-continental Railways were built.

The Santa Fe Railway followed the route of the old Trail and when it was completed, this famous and glamorous old Trail was no longer used and so closed one of the chapters of the story of the old West. A long chapter filled with the deeds of courageous men who daily dared death in their long treks across the wilderness, not so much for the gains of the journey, as for the sheer love of high adventure, the lure of the unknown and the joy of life in the open. Many pages of this chapter were stained with blood, over the price that men have paid for conquering the wilderness.

When the days of this old Trail were ended the men who had tasted of its enchantments were forced to seek their excitements and adventures elsewhere. Some remained in the Southwestern points, Santa Fe, Taos and others of the Mexican towns where dark-eyed Senoritas had ever smiled on the Americans. Many went on to the Pacific Coast states. Many others found the hectic life they craved in the Kansas towns where the Texas cattle drives centered.

In 1877, in Dodge City, there were numbers of these old Santa Fe Trail men. One of these, with whom I was closely associated in later years, was Sam Kyger. He took up cattle ranching in the country, one of the first of the ranchers south of Dodge. His experience of trading with the Mexicans was not a pleasant memory.

Sam, in company with a partner whose name I do not recall, left Missouri in the later days of the Trail, when the railway had reached Granada, Colorado, where it stopped for a time. They shipped a carload of fine stallions out to Granada and then drove them overland to New Mexico for trade and barter. Well within the borders, in a wild and desolate place one night, they were set upon by robbers, their fine animals stolen and they, themselves, were forced to flee for their lives. Mr. Kyger's companion was wounded and in their flight in the darkness, amid unfamiliar surroundings they became separated. Kyger made his way out to safety, but he never saw or heard of his friend again. He appealed to the American Council at San Luis Potosi, the Port of Entry, and search was made for his companion, but to no avail.

Mr. Kyger then settled down to cattle ranching, as a better and safer business than trading with the Mexicans, and he continued in it until his death many years later. Kyger Creek in Clark County bears his name.

After the Santa Fe Railway was built into Dodge City and that town became a main outfitting point, a number of North and South Trails developed. Lesser ones it is true than the great Santa Fe Trail, but in their own way and time, just as important to the development of the country they served.

Part 2 • Chapter 2

In 1868, a trail was made from Fort Dodge on the Arkansas River to Camp Supply, at the junction of the Beaver and Wolf Creek. This Camp of Supply, afterward known as Camp Supply, was established as a base of supplies for white troops under General Custer, making a winter campaign against the Indians as punishment for raids and forays against white settlers.

It was necessary to rush these supplies to the troops before winter set in so a train of one hundred wagons, drawn by six hundred mules, was sent from Ft. Hays to Dodge City, over an established trail between these Forts. But from Ft. Dodge the way led over the unbroken prairie south across Mulberry Creek and on to Bluff Creek and on to where Ashland, Kansas is now, then on to the Cimarron River, crossing it about where the old Texas cattle or Through Trail later was and so on to Camp Supply. This wagon train made two trips with freight over this way in the fall of 1868, and so was established the first North to South trail in this part of the West.

When the Santa Fe Railroad, built into Dodge City, in 1872, that town became the outfitting point for all the Government Forts to the South and Southwest. Soon, too, a large part of the Texas cattle movement was directed to it and this explains somewhat the growth, importance and prosperity of the town at that time.

The Adobe Walls Trail was made in 1874. By that time the buffalo had been driven far to the South and West and the hunting in the country adjacent to Dodge City was no longer profitable. A number of the hunters agreed that Old Adobe Walls was a better base for their operations and Dodge City merchants and outfitters put up buildings and hauled in supplies, these for the convenience of the hunters and the buffalo hides would then be freighted to Dodge City for shipment. This trail was made but barely had these arrangements been completed when the hunters, store keepers and all, numbering 28 men and one woman were surprised and attacked by a large force of hostile Indians, thus followed the historic Adobe Walls fight, in which almost miraculously, the whites escaped extermination.

Twenty-eight men against something like 1000 Indians, it was a thrilling event. The story of this fight was told to me a few years later from a number of the participants, for I became well acquainted with most of the men who were in this fight.

I do not know the exact time of the establishment of the other trails I shall describe, but they were made in the next few years as the need for them arose.

These trails were not surveyed but their way was picked out by the plainsmen, freighters, and cattlemen. These men selected the way well for always these trails led over the way that would be best adapted to all times and seasons, wet or dry, and with due regard for convenient watering places at regular intervals, the best crossings of the streams to be forded, and always keeping the general direction toward the objective point.

At the time of my coming to this country there was, leading out from Dodge City to the South, just one very wide, smooth, hard beaten trail. About five miles from the town this trail branched. The branch that led almost directly South was the one that extended to the coast country of Southern Texas and it was then called "The Through Trail." Over it were driven the tens of thousands of cattle that gave Dodge City its title "The Cowboy Capital of the Southwest." These Texas cattle were then called "through cattle" meaning they had come from Southern Texas.

South of the Cimarron River, a branch trail left the Through Trail and led in a southeasterly direction to Camp Supply and from there on to Ft. Elliott. This was called the Supply Trail.

The Tuttle Trail, a cattle and freight trail, was another branch of the Through Trail that left it some seven or eight miles south of Dodge City and this trail led to the Tuttle Ranch on the Canadian River in Texas.

The other branch of the trail, five miles out from Dodge City, bore more to the southwest, and was called the Jones and Plummer Trail. It led to the Jones and Plummer Ranch on the Canadian River.

The Tascosa Trail was another branch of the Jones & Plummer, which left it a few miles south of the present Beaver City, going southwest to the head of Palo Duro Creek from thence to the Canadian River and on to Tascosa, Texas.

At a later date the Healy Trail was added to these earlier ones. Over these trails by slow moving wagon trains, drawn by patient oxen or hardy Missouri mules, carried the vast store of supplies that sustained life in distant Army out-posts and lonely isolated ranches scattered from far to the south and west of the one point of supply that was Dodge City.

The retail trade of the main outfitting stores was enormous. Often their supplies came in not by the car-load, but by train-loads.

York & Drapers, Wright and Beverly's were the main outfitting stores in 1877, a few years later the Geo. Emerson Co. Store was established. As there was no bank in the town, all of these stores carried a small banking business as a sideline.

Part 2 • Chapter 2

In consequence of this enormous retail business the freighting industry was a most important one. The price of freighting was from a cent to one and one half cent per hundred pounds to the mile.

Among the well known freighters of that day were Leo Winer, Archie Keech, Lee Harlan, Frank Stafford, "Old Man" York, The Coombe Brothers, and many others I could mention, all masters of transportation of those days. These men saw many hardships, fought Indians, blizzards, quick sand and high water in fording streams.

The largest freighting outfits of that time were owned by Lee and Reynolds or Roth and Wright, though there were many smaller outfits. These wagon trains sometimes had as many as forty or fifty wagons in a train, each wagon with one and sometimes two trail wagons, each load wagon with from six to ten yoke of oxen or six, eight or ten mules as the case might be.

Oxen were used to a great extent in the summer time for they could live on the buffalo grass but in winter the oxen were replaced by mules and these were fed grain and hay. Each wagon train had a wagon boss and his word was law. Sometimes the oxen became tender-footed and had to be shod while on the trail. Every "bull-whacker" could shoe an ox and every "mule skinner" had to be able to shoe a mule if necessary while out on the trail. These men were a hardy lot in the prime of life and the weeks of living in the open gave a zest to life on the Plains, an excitement in the blood that dwarfed all the hardships.

The long trips on the trail to far distant points, though seemingly continuous, were never without incidents and happenings that kept the men constantly on their guard. And when the long trek over, the days in town were carefree and happy ones.

Cimarron Chronicles

Chapter III

The Texas Cattle Drives

In 1878, and several following years, great numbers of cattle, thousands upon thousands of them, were driven up from Texas to Dodge City.

During the Civil War, Texas had been practically isolated from the rest of the United States and had become greatly overstocked. And so began the cattle drives to the Railroads, lying along the Southern part of Kansas, starting first by way of the Chisholm Trail from the Red River in Texas to Abilene, Kansas. As the railways were built on West, the cattle drives shifted to the west and so at this time Dodge City was perhaps the most important of all the shipping points.

Many of the herds at that time were shipped or sold at Dodge City. Many more were taken on north as far as Montana; for all that northern country was one vast un-trampled range and now was being stocked with Texas cattle.

The Texas ranchmen soon learned that Texas steers moved to a northern climate would, at maturity, weigh at least three hundred pounds more than when matured in his native state. So now many of the large ranchers of the Texas Country were establishing branch ranches in Montana and other northern points. From the home ranch in Texas each year a herd of young stock was driven to the Northern ranch for maturity and in time from thence were sent to Eastern beef markets.

These "Through Herds," as they were then called, generally consisted of from 2500 to 3000 head of steers or heifers. The drive started in Texas, sometimes in February, and proceeded north as the coming of green grass permitted. Usually the first of these herds reached Dodge City about May 20th, the full trip to Montana being completed in from six to seven months. As Dodge City being the first town on the trail after leaving Texas, there was always a stop there for rest, recreation and whoopee.

As my first work was with some of these Texas herds, I made the acquaintance of a number of these boys who came up from that country with the "through herds." I found them almost without exception, fine fellows, generous, sociable, good company, good cowhands, but a great many of them could neither read nor write. For in the years following the Civil War, Texas had almost no rural schools and these boys brought up in the "Cow" country had no chance for schooling.

Some of these boys reared in the "brush" country of southern Texas had never seen a railroad train until they chanced to come north with a herd. One of them once told me of his reaction at his first sight of a moving train. It was at Dallas, sitting on his horse that he heard a whistle and shrieks and saw a train come rolling in. Sitting there, he watched in amazement its swift gliding approach. Suddenly the thought occurred to him that as long as the thing was coming toward him in a long line like that, he could easily keep out of its way but what if it should come at him broadside? Panic seized him at the thought and applying the spurs to his horse, he fled madly never stopping until a quarter of a mile away. Then looking back he saw that the train was far past and still gliding smoothly along those gleaming rails. It had never occurred to him that the train could not leave the track and go right across the prairie just the same as a horse and in telling me the joke on himself he could laugh about it just as heartily as I did. That was one of the likeable traits of those boys, they were always so frank to acknowledge their greenness.

Life on the trail, from Texas to Montana, was both strenuous and monotonous for these boys, but some of them made this trip a number of times. Besides driving the cattle by day, they had to be guarded by night.

After the herd was bedded down, the men were divided into shifts of two to stand guard for nearly two and one half hours for each shift, except in case of rain or storm when sometimes all hands had to turn out to help avoid or participate in a stampede, and those occurred at times seemingly without reason. The herd might be resting quietly on the ground when suddenly with a snort of fear and alarm one animal would bolt. Instantly, as if an electric current had passed through them all, the whole herd was up and away, panic-stricken. The riders must race with them, striving to turn back the leaders so that finally the whole bunch milled in a circle for that was the only way a stampede could be checked. And woe to the rider whose horse fell with him in this mad race in the dark, over rough unfamiliar ground, through prairie dog towns, or wherever the race led.

Part 2 • Chapter 3

Some nights, the herd would be nervous and more than one stampede took place. At each time, when the cattle seemed nervous, the boys would sing to them as they circled the herd and the sound of human voices seemed to give the cattle confidence, soothed and quieted them. The morning after a stampede the herd had to be counted, and if any were missing they must be found and returned to their place before the herd could move on.

At times several herds camped at night within a few miles of each other and then if a stampede occurred there was a grand mix-up. Then each outfit had to cut out the cattle of their brands and get all straightened out for the trail once more. Every herd traveled under a road brand.

These Texas outfits generally had a Negro cook with a chuck wagon. These Negroes were desperately afraid of Indians and coming through the Indian Nation many laughable incidents took place. For, at times, the wagon got far behind the herd. That was the time selected by the young braves to have some fun and get into mischief. Usually they were merely looking for food, but in either event the darky, scared out of his wits, applied the whip and dashed madly in pursuit of the herd, never stopping until well within the bunch, thereby causing another stampede.

The food on the long trail north was very plain fare, coffee, salt pork, corn pone and molasses, every day three times a day. White bread was a luxury. Occasionally, of course, there was fresh beef to break the monotony of the daily fare.

After several months of this life on the trail it is no wonder that the boys when they reached Dodge City were ready for anything. Here, many of the herds were sold and the boys paid off, and most of them spent every penny they possessed before leaving town. So it was rather fortunate that it was customary for the outfit they worked for to provide transportation back home.

The first cattle that were held on the Crooked Creek and the Cimarron River, west of the Through Trail, was one of those Texas herd's that came in late in the fall of 1877. They were owned by a man named Lovell. Coming in here with five thousand steers and finding the grass so fine and abundant, it was decided to winter them here. They were turned loose and kept within certain bounds by means of line riders. Line camps were then established at different widely separated points along the Cimarron River and Crooked Creek valleys.

At these line camps usually two men were stationed. Each morning they rode in opposite directions meeting other riders at a given point, all turning in the cattle along the line they rode, and so each day making a complete circle. The cattle were easily kept within prescribed bounds.

The Lovell outfit was not afraid of work for at each of their line camps they made roomy, comfortable, if somewhat crude, dugouts with the materials at hand. No windows in these, but a fireplace at the back furnished both light and warmth for the long winter evenings.

Their winter supplies were brought out from Dodge City and the following season the Lovell cattle were taken on north. The camp dugouts remained as landmarks and were utilized by the cattlemen that came into the country soon afterward.

In the fall of 1877, George and Hi Kollar settled on the piece known as the 76 Ranch, northeast of the present site of Ashland, Kansas. The same fall John Frazier settled about three miles north of where Protection, Kansas, now is, on Kiowa Creek, and built a log house there. His cattle ranged on Bluff, Kiowa and Cavalry Creeks, his range on Bluff Creek adjoining that of Hi Kollar's. At the mouth of Bluff, Henry Kollar, a brother of Hi and George, established his ranch, ranged up Bluff and joined Frazier's.

At that time these were the only ranches on the three Creeks named, but that same fall Sam Kyger settled on the creek northwest of Ashland that still bears the name of Kyger Creek. It was in connection with this isolated, lonely ranch, that Mr. Kyger later told us a story that was a strange blend of the tragic and comic.

It was in early spring of 1878, that he brought to his ranch from Dodge City a family fresh from Arkansas, a father, mother and two half grown boys. In a short time the mother sickened and died. Mr. Kyger was away at the time so it devolved upon the father to go to Dodge City for a casket that had to be made while he waited. Leaving the two young boys to guard the corpse, the father on leaving gave them his cherished fiddle to help pass the time away and at the same time giving them strict orders to keep the cats out of the house for Mr. Kyger was fond of cats and always kept several of them around the place.

The boys, however, became careless and the cats did get in the house and mutilated the face of the corpse. Mr. Kyger returned home just before the return of the father with the coffin, and this, according to his story, was how the father took his sons to task for their carelessness: "There you sat, see-saw, fiddle dee de, playing on that fiddle and let the cats eat on your mother's face. Now look at that, a pretty corpse isn't it? Why, it ain't fit to go to hell with." They buried the mother not far from the dugout and father and sons then left the country.

Part 2 • Chapter 3

In 1878, ranches began to be established west of the Through Trail. Driscoll settled east of the trail and ranged from the south of Bluff Creek west to the trail. Beverly Brothers settled on Sand Creek, southwest of Ashland, at Blue Moles and ranged from Englewood east to the Through Trail and north of the sand hills.

In Meade County far up on Crooked Creek, in what is now a part of the Artesian Valley, were a few settlers, the very first of a lot of pioneers that arrived the following year.

I was up on the Buckner, north of Dodge City that summer, working with a through herd and here, too, settlers now began to appear. I remember a young man started a store there that summer on, or very near, the present site of Jetmore. So, at this early date, appeared the vanguard at widely separated points, of the settlers and cattlemen who came in such numbers in the following years. The settlers came to make homes in the West, the cattlemen came because they saw here immense possibilities, almost unlimited prospects for the future in the cattle business.

If, at that time, anyone had predicted that in ten short years all this vast virgin land could become crowded and overstocked, they would have been hooted at. But such, never the less, was the case.

In Camp

Two old time cow punchers, "Medicine" Ben Steadman and Alex Crawford.

Part 2 • Chapter 3

Two old cow punchers, Bill Wright and Sebe Jones.

Two old cow punchers, "Doc" Anshutz and "Irish" McGovern.

Chapter IV

Indians

While I was still in Dodge City, sometime in September 1878, the northern Cheyenne Indians dissatisfied with their new home in the Indian Territory, gave the soldiers the slip and make a dash for the north to return to their old home.

Coming through this country a little west of the mouth of Bluff Creek, they came upon a part of the Driscoll outfit building line camps for winter line riders.

The wagon was camped at a water hole about two miles from where the men were making a little dug out on a creek. A band of young warriors out on a foraging trip, away from the main band, came upon the wagon and demanded food. The cook had a pot of beans on the fire, which he prepared to serve them. The young bucks feeling very gay began to use their quoits on the cook and otherwise mistreated him, arousing his ire. He quietly slipped a small portion of strychnine in the bean pot, which he then set out for his tormentors.

Very soon after partaking of these, the Indians began falling into convulsions and during the excitement that ensued the cook ran to where the Indians' horses were standing and jumping on one of them made his way to where the men were working. Returning to the wagon with him they found several dead warriors, while the remainder fled in terror. The dead were buried in a wash out. This incident was not generally known for some years because of the Governments objection to killing Indians.

Another small band of young braves found the dugout that was on the Driscoll home ranch. Jess Driscoll was alone here and when the band began to plunder the place he fired on them, killing one of the number and making it so hot for the others that they could not pick up and carry off their dead as was their custom. He afterward went out and scalped the young brave, hung the scalp on the dugout door to dry and afterward displayed this gruesome token with much pride.

Sam Williams also had a brush with the Indians. Sam was a hard-boiled old timer, a sheep man; he had ranched in New Mexico for some time and had now moved to this country. He was taking his sheep north to market and had reached Sand Creek in southwestern Meade County when a band of young warriors came upon him and his herder, a young Mexican. They were in camp at the wagon when the Indian band appeared. Jumping on their horses, they raced away with part of the Indians in hot pursuit, while the remainder began slaughtering sheep.

Suddenly the horse, ridden by the Mexican boy, stumbled and fell, throwing his rider. With a terrified cry the boy arose but the horse was up and away. Mr. Williams, without slacking his speed for an instant, turned and circled the boy. As he rode in diminishing circles, the boy, when he came in close enough, made a desperate leap and landed on the back of the gallant little sorrel mare ridden by Mr. Williams. Now carrying double, the little sorrel mare distinguished herself by completely outrunning the pursuers. It was a thrilling race, but she carried them to safety and ever after this the little sorrel mare "Nellie" was the pet and pride of her owner. Several hundred head of the sheep were slaughtered, the rest scattered.

With Dodge City as their objective, after leaving the Indians far behind, Mr. Williams and the boy went on their way. They had not gone far when they met some of the cattlemen of the country, for all had heard of the invasion and were now banded together and prepared to give fight to the enemy. It was remarkable the short time it took for such news to travel, but the Indians had sent out small foraging parties all over the country as they had left their reservation hastily and by stealth, without much in the way of supplies.

Now these ranchmen were gathered with the avowed purpose of exterminating the Indian band. Corralling them one evening on a little creek, east of the present site of Meade, Kansas, they were preparing for the attack, but were prevented by the arrival of soldiers from Ft. Dodge. All were eager for the fight, but the officers forbade any action, so, in deep disgust the cattlemen returned to their scattered ranches and the sheep men to their scattered herds, for another band of sheep besides Mr. William's, also on their way north, had been attacked and suffered quite a loss. Next morning the soldiers found that the Indians had slipped quietly away during the night.

From Sand Creek they went northwest, next encountering the few settlers on Crooked Creek, in what is now the Artesian country in Meade County. Here these settlers had various experiences, some leaving their goods and chattels that lured the Indians.

Part 2 • Chapter 4

At Captain French's place they found a grindstone and forced the Captain to grind their knives for them. They did not otherwise molest him and the other men who were with him, realizing, no doubt, that they were pressed for time, as the soldiers were on their trail.

But a few miles farther on, along the Adobe Walls Trail, they killed a freighter and looted his wagons; here was a chance for acquiring more of the needed supplies.

They then proceeded north across Crooked Creek and on to the Arkansas River, crossing it about four miles west of Cimarron, Kansas, which then consisted of a section house and not much else. It was near here that I had a sight of the Indian band. I had been up the river delivering a letter to the boss of a through herd, and was on my way back when I encountered them. They fired a few shots in my direction, probably just to scare me and in this they succeeded admirably for I made the section house on a wild run.

They left one decrepit old squaw dead on the Arkansas River, for they were being hard pressed and bearing ever in a northwestern direction and moving rapidly, they managed to keep ahead of the soldiers.

After getting farther on, however, they engaged in several running battles with the soldiers in which numbers of the Indians were killed so that only a remnant of the tribe finally reached the old home in the north. Here they were overtaken and in a terrible battle in which the Red Men were almost annihilated they were forced to surrender and the captives were returned south, their fierce war-like spirits broken and subdued forever. Eight of the leaders were brought to Dodge City and kept in the jail there for some months, but these too were eventually returned to the "Indian Territory."

Some year later I saw one of these old Chiefs at Camp Supply and remarked to Joe King, who was with me, "I have seen this fellow somewhere before." The old fellow heard the remark and crossing his finger like # to indicate bars, he placed them before his eyes and peered through at the same time, grunting, "Dodge," meaning that I had seen him through the bars at Dodge City, for all the town went to see the captives while they were there.

While I am on the subject, I may as well tell of another Indian scare. In the summer of 1877, Doc Burton settled on the head of Wolf Creek in Texas. At the same time Tom Connell settled fourteen miles lower down on Wolf Creek

As usual dugouts were their living quarters and on the Burton place were two, one for dwelling and the other for a storehouse. As no lumber was available, even the door frames were made of hewn cottonwood and a blanket hung over the opening served as a door. At times roving bands of Indians passed through on hunting expeditions. One day Ed, the Negro cook, alone on the place, saw two Indians ride up. Ed lived in mortal terror of the "Red Men" and he ran to the storehouse to hide.

Outside at the top of the steps stood a grindstone and Ed peeping out saw his visitors grinding their knives. Presently, one of them decided to investigate the storehouse and Ed, to use his own word "scrooched" down behind a box in the darkest corner, above him some meat hung, suspended from the ridgepole. The Indian, looking about with prying eyes, soon discovered the meat and with eyes upraised walked over and whipped out his long knife to cut it down. The sight of that knife was too much for the crouching Negro, who now sprang up from under the very feet of the Indian with the single exclamation of "oop!"

This apparition of black face and rolling eyes showing mostly white, rising seemingly from the very ground beneath him, was too much for the Indian, even more terrified than the Negro. Both reached the doorway at the same instant, which resulted that they carried the doorframe with them to the top of the steps and here at opposite ends of the frame each trying to get away from the other. The Indian fell down and Ed, slipping from under the frame, took off down the creek in the general direction of the Connell Ranch fourteen miles away.

Doc Burton, riding up a few minutes later chased the Indians away and not finding Ed anywhere about, surmised that he had started for the neighboring ranch. Thinking that he could overtake him soon, he started in pursuit and though he rode at good speed all the way to the Connell Ranch, he found that Ed, with fear-winged feet, had beaten him to it.

Part 2 • Chapter 4

Outfit branding on the prairie on Palo Duro Creek.

Cimarron Chronicles

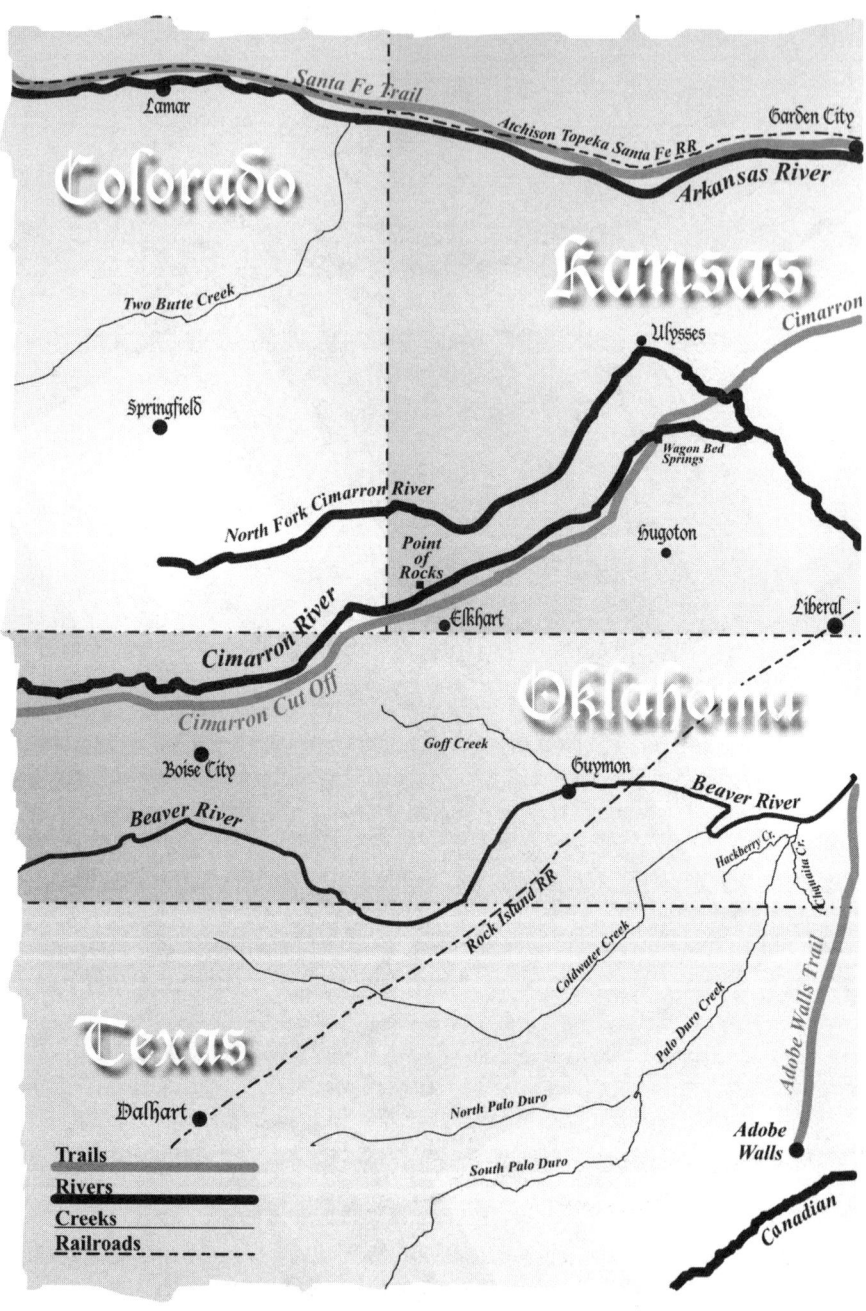

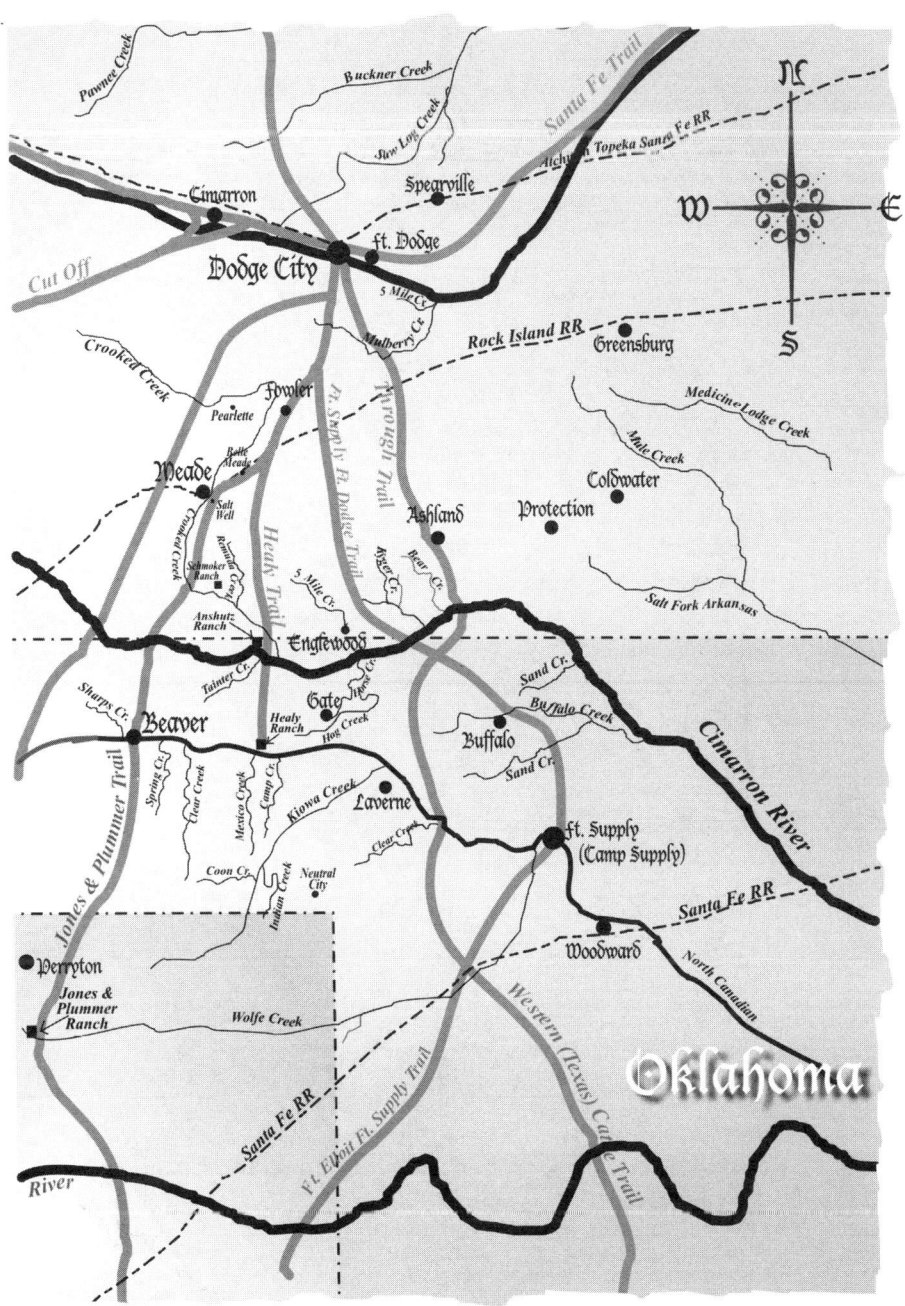

Examining a disputed brand.

Chapter V

The Ranchers

The Burton and Connell ranches were the only ones on Wolf Creek at that time, but in 1878, other ranchers settled there, among these were George Anderson, who started the "X" brand. Later he sold out there and moved up to this country.

The winter of 1878, was a hard one up on the Arkansas River and the country north of Dodge City. Cattle from the Smoky River drifted down and filled the cuts along the railroad so that the trains killed numbers of them and it became necessary to have riders to clear the cuts of cattle.

For a time I helped with this work and it was that winter that I first saw snow plows used, but they were crude and clumsy compared with the ones in use at the present time.

A good many cattle and sheep perished that winter, as well as one man wintering his sheep. Another man wintering his sheep on Mulberry Creek, south of Dodge City, lost his entire flock, for the snow was very deep and he had no shelter for them whatever.

I experienced my first blizzard that winter. We could hear it coming across the great plain to the north, a roaring noise. The storm raged for thirty hours with great fury.

In the spring of 1879, Judge Beverly, of Dodge City, sent another man and myself to the Cheyenne Bottoms, at Great Bend, to get a bunch of horses that had been wintered there. These horses had belonged with a through herd that was sold in Dodge City the previous autumn. Not finding any sale for the horses at that time they were sent away for the winter.

When we returned to Dodge City with them, the Judge left them in my care to be sold as opportunity offered. I grazed them along the river bottom below the stockyards for some time. Some of them had wintered well, others were in poor condition.

One morning a very cocky old Dutchman came to look at the horses, he walked down through the herd, carrying a closed umbrella under his arm. After looking them over he came up to me and said, "Do you make horses here?" I said, "No, why?" "Oh," he replied, "I see you have some good frames here."

After this pleasantry, he was all set for trading and in due time, I sold him a couple of gentle old ponies, and a saddle, bridle and blanket. We rode the horses for him, and then he got on and tried them and was pleased with his purchases. As he prepared to leave, he mounted the extra gentle one, intending to lead the other, but before gathering up the lead rope, the sun having come out hot, he shot up the umbrella. This was an indignity probably never perpetrated on a Texas cow pony.

The raising of that umbrella over his hurricane deck brought instant action. Buckling up in the center, he proceeded to do both plain and fancy pitching, while the Dutchman yelled mightily, soon landing on the ground in the midst of his ruined umbrella. I walked up and said to him, "Do you make umbrellas?" "No Vy", was his reply. "Oh", I replied, "I see you have the frame for one." "Oh, now," said he, "You dink you are so smart." And he rode off in a huff.

The summer of 1879, saw the first movement of cattle ranching on the Cimarron and Beaver Rivers, west of the Through Trail.

Old man Choate, everyone called him that, brought his cattle from up on the Pawnee in Kansas, for the settlers were driving the cattlemen out up there. Choate ranged from the Through Trail up the Cimarron River to Horse Creek. Sam Bullard and the Fitzgerald Brothers, Herb and Charley, all Boston men, brought up a large herd from Texas and located south of the Cimarron River on what is now known as Taintor Creek. Here they built a house, half dugout and half log house, near the mouth of the creek. Their range was from Horse Creek to about twelve miles west on the Cimarron River, also north along Crooked Creek.

In 1881, Bullard sold his cowherd to Fred Taintor and then established another ranch close to Garden City.

R. K. Perry, who also came down from the Pawnee in 1879, was the next ranchman up the Cimarron River. His was the "T" Ranch, with him came N. J. Rhodes and his son Wiley, then a mere boy, and settled nearby. Several years later Mr. Rhodes moved his family to the ranch and Mr. Perry married and brought his wife to the ranch also.

Beyond them was the ϙ Egg Half Diamond outfit owned by a Mr. Deardoff of Kansas City. His herd was brought in from the Arkansas River, west of Dodge City. The cattle were in charge of "Old Man" Holloway, an odd and humorous character.

Joining the egg half-diamond range on the west was the range of the O'Neil Brothers, Hugh, Henry, Paddy and Jim. As their name would indicate, these boys hailed from Old Ireland. A few years later when they sold this ranch to the McCoy outfit, Hugh went to Ireland, married the girl of his choice and returned bringing back a wife for Henry, also.

On the Beaver River the R-S, a Texas outfit, ranged from Camp Supply west to the Through Trail and north on Buffalo Creek. West of them, a man named Cox, also from Texas, ranged his cattle from the old trail west to the Kiowa. On Camp Creek a tributary of Kiowa was another Texan named Catlin. There were others up the Kiowa, whose names I do not recall, but John Over, of the M Ranch held a range from the mouth of the Kiowa to the mouth of Mexico Creek. Healy Brothers joined him on the west and their range extended from the mouth of Mexico to the mouth of Clear Creek east of the present site of Beaver City.

The freighting industry from Dodge City to many distant points in Texas was an important one at this time, so in 1879 and 1880, a number of road ranches were established for the accommodation of freighters, these were usually at the places where the trails crossed streams. At Crooked Creek just east of where Meade now is, "Hoo-Doo" Brown put up sod buildings, one for a dwelling and the other a store. At the Cimarron Crossing was Charley Heinz place and at the Beaver River a road ranch was built by Nola Carey and Pete Lane. All of these were on the Jones & Plummer Trail.

At Sharp's Creek on the Beaver River, a man named Crawford, built a road ranch on the Adobe Walls Trail and where that trail crossed the Cimarron River, three brothers, Frenchmen, had a place. They had not been in this country very long and could scarcely speak the language. One day I was at their place when my attention was attracted by numbers of buzzards that were flying about and I remarked to one of the brothers, "Joe, why don't you kill some of those buzzards?' "Oh," he answered, "I boil a buzzard, I fry a buzzard and I bake a buzzard and buzzard no good any way I cook him."

At these road ranches meals were furnished. Also, one might buy provisions, tobacco and liquor. In winter, hay was available for the freighters mules. P.G. Reynolds had contracts for all mail and stage routes out of Dodge City to the south and southwest into Texas. On some of these lines the mail was carried in buckboards but on the route from Dodge City to Camp Supply and from there to Fort Elliott the regular old style stage coaches were used. In order to make good time on these mail and stage routes a change of horses was required at every twenty-five or thirty miles. On the mail route over the Jones & Plummer Trail these stage horses were kept at road ranches or at the place where a post office might be established.

On the Through Trail, Charley Coffman came to Deep Hole, south of Ashland, Kansas, in Clark County, Kansas in 1879, and built a farmhouse for a store and a road ranch. Previous to his coming, a small detail of soldiers were stationed here where they had a sod house and a corral also of sod. These soldiers were kept at this outpost for general peace purposes and to protect property. Here P.G. Reynolds always kept a change of horses. After Coffman's coming a post office was established, I think it was called "Deep Hole" and the ranchers for many miles around came here for their mail.

The cowmen kept a man here whose duty it was to inspect all herds coming up the trail for strays that might have been picked up along the way. A man was also kept in Dodge City for this purpose. The Coffman place did a thriving business in selling supplies, whiskey and tobacco to the ranchers, trail men and other travelers many of whom stopped here overnight.

There were a couple of other road ranches on this trail at that time, one several miles north of the Cimarron River near Bear Creek. This was a picket house, and another farther south of the Buffalo River on the Supply Trail. This was also a picket house and was kept by a man called "Brushy Bill," that is the only name I ever heard for him, a nick-name, for his hair and whiskers were an unkept mass of nondescript color that looked exactly like a bunch of brush; from appearances, I would say he had never used a comb.

On the head of Bluff Creek was another ranch where P. G. Reynolds kept a change of horses.

Until the advent of the settlers, and the towns and railroads that followed, the Reynolds mail and stage lines faithfully served the pioneer cattlemen and first settlers of the long trails from Dodge City to distant points in Texas.

Chapter VI

Adventures

The summer of 1880, I spent in Colorado going from Dodge City to Pueblo. It had been a dry season and the cattle ranchers in eastern Colorado had no grass for their herds. I worked at what I could find to do. A little cattle herding and some freighting and that brought me a new experience, that of driving a six mule team, with jerk line, over one of the passes into the Alamosa country. A jerk line, I should explain, was a single line run through home rings the entire length of the teams and fastened to the leaders; they kept apart by a jockey stick. A pull on the line for "Whoa" and a jerk on the line for "Go," the well-trained leaders responded promptly to this simple guidance and the others of necessity followed.

It was something of a marvel how quickly the lead team would respond to this pull and jerk and the swing teams had to be equally alert to follow their leaders or they would be swept off their feet by the swing of the heavy chains.

The driver always rode the off wheel mule. The harness was free of any unnecessary trappings for there were no throat latches on the bridles, no bellybands, only back bands on the harness, and no buckles on the collars. When unharnessing the driver had only to unbuckle the ham strap, peel back the harness, pull off the bridle and the mule was free.

All freight outfits using six or more mules used the jerk line and the lack of buckles and straps made the job of hitching and unhitching such a number of animals a speedy affair.

In the mountains, all teams were required to wear bells on their harness for the roads over the mountain passes at that time were narrow, rough and dangerous in places. At regular intervals along the pass, the road widened to permit passing in opposite directions. On reaching these places the drivers were required to stop their teams and listen for the sound of bells, if none were heard one might go on.

Even so it sometimes happened that teams did meet on the narrow parts of the trail and then followed trouble, serious trouble indeed, sometimes even fights to settle the disputes that arose as to which outfit would back up to one of the passing places.

However, we had the good fortune to make our way over this pass without any such unpleasant experiences. Back from that trip, I helped freight cross ties out from the Greenhorn Mountain side and hauled to the Denver and Rio Grande R. R. on the Huerfano.

Many of these ties were cut so far up on the mountainside that chutes were made to carry them down to the loading places. On certain days the chutes were used and it was a sight to see the ties come hurtling down the chute at terrific speed and at the end shoot off into space for several hundred feet. Dangerous too, it was to be in the vicinity.

On one trip to the Huerfano with a load of ties we were caught in a cloudburst. As it happened we were on a little flat, a small level space and when the rain ceased the water there on that level place was up to the wagon reach. I did not know rain could fall so fast and in such quantities, so could scarcely breathe. My employer, a Tennessean, named Nimrod Miller, was adept at profanity, and on that occasion I think he quite exhausted his vocabulary.

The wind blew almost incessantly that summer up there on the eastern slope of the Rockies. It sometimes reached such violence that large pebbles were blown in one's face and it was almost impossible to work out in the gales.

On the 4th of July, I was high up on the mountain and that night it snowed a heavy fall that extended clear to the foothills. I decided right then and there that I would change my location at the first opportunity, and this was soon presented for I got a job with a cow outfit to help move a herd to eastern Colorado, near the Kansas line.

On the trip down we had a stampede near Bents Old Fort on the north side of the Arkansas River. It was a very dark night and while riding at full speed after the stampeding herd, one of the boys had the misfortune of running into a tree cactus that raked one side and filled him with cactus spines and thorns of all sizes. He was in a pitiable condition. We got him to camp and when daylight came the boss assigned me to the task of removing the cactus thorns. This was no small job and was accomplished by the aid of a pocketknife and amid wild yells and much cursing. When the last thorn had been removed I bathed him in warm water and anointed the irritated parts with some of the cook's lard, for that was all we had in the way of ointment. For

Part 2 • Chapter 6

several days thereafter, he was obliged to ride on the wagon with the cook, and for an old time cowpuncher that was adding insult to injury.

Our way led past New Fort Lyon, the old Fort had been abandoned some time previously and the troops moved to the new fort. We went on to the old Fort, some miles below, and finding fine and abundant grazing there, decided to lay over for a half-day and let the cattle fill up.

We were all sitting around on the crumbling stonewalls of the old Fort when about 3 P.M. a little cloud appeared. An hour later it looked wicked, rain began to fall, and a few large hailstones came spattering down. We all mounted our horses to go to the herd and I picked up an old wash boiler lying there and placed it over my head to protect me from the hail that was soon coming down thick and fast.

The cattle stampeded and our horses, frantic under the pounding hailstones, did likewise. I had all I could do to keep my seat in the saddle, for the noise of the hail on my boiler did not tend to quiet my mount. When the hailstorm was over, my boiler was all dented up out of shape, but I was unhurt, the other boys without protection were badly beaten up. It took us two days to gather up our scattered herd and get under way once more.

When we reached our destination, about August 10th, I returned to Dodge City.

It was only a short time later that I met Shangh Pierce, a tall rangy Texan, with a booming voice that matched his big body. He had a herd on the trail coming up from Texas and had reached Dodge City ahead of it. He now wished to go out to meet the herd and find how they were coming on, so hiring a livery rig and myself as driver, we started out over the Through Trail. Somewhere in the near vicinity of the present town of Ashland, Kansas we came upon the herd and drove into camp. Only the Negro cook was at the wagon, a still form covered by a wagon sheet was nearby on the ground.

The boys, the cook explained, had just been having a horse race, an argument and quarrel ensued in which this man was shot and killed. We buried him close by the next day. The killer got on his horse and rode off, no one knew where to, but he was no longer welcome to stay with the outfit, and that closed the incident. Such occurrences were not uncommon in these days. This type of killing, wherein the man quickest on the draw won the argument, was rather lightly passed over, and was accepted as a way to end quarrels.

However, there were crimes committed in those days that were not so lightly condoned. These embraced premeditated murder and robbery, things of evil planning and black mystery.

Such seemed to have been the fate of two young men who came out from the East for hunting and adventure. They came first to Dodge City and later went to Colorado. At Trinidad they bought team, wagon and equipment for camping and hunting. Before leaving town they wrote relatives back home of their intentions and then struck off into the wild to the south.

They were never heard of again, their relatives came from the East and much searching was made, but no trace was ever found of these two unfortunates and the outfit they started out with. The supposition was that they were murdered and robbed.

But another case, more widely known, was one that occurred at one of the earliest established ranches in this part of the West. After settling on this place that was a wild and isolated one, with a great stretch of range, the ranch owners made a trip to New Mexico and purchased, or rather bargained for, a herd of cattle to be delivered at the ranch.

The owner of these cattle was a very wealthy man of Spanish Mexican lineage, an influential man of his time and place. According to the agreement the cattle were gathered and trailed to the ranch of the purchasers far to the northeast. In charge of the herd was the young son, heir of the owner to whom payment was to be made on delivery of the cattle.

In due time the herd arrived and was turned over to the purchasing parties. What transpired then was never fully known, only a few bare facts were ever fully established and these were that the young man had instructed his men to start back home with the wagon and horses and that he would catch up with them in a few hours. There yet remained a few details of the payment to be finished before he could leave.

The Mexicans started on their way traveling slowly. In vain they watched the receding horizon for a sight of the young master on his spirited horse, decked out in saddle and bridle of rich Mexican handwork as befitted one of his rank in life. The long days dragged on, at last the New Mexico home ranch was reached and report was made.

The father with a number of men at once took up the search. Returning to the ranch, where the cattle had been delivered, he was courteously received and assured that his son had received cash payment for the cattle and gone on his way, further than that, they knew nothing. The grief stricken father was confident that there had been foul play and spent weeks in the surrounding area with his men, searching every foot of ground for a clue. They investigated every possible grave site; they dug all over a vast scope of country in search of the body. However all their work uncovered not one single clue to the mystery and at last in sorrow and despair the search was abandoned and the son, the cherished hope of a fond family, was never seen nor heard of again.

It was this dark and mysterious disappearance that was often times discussed and speculated on around our campfires. In lonely dugout line camps, lighted only by the flames of the fireplace in the long winter evenings, this story was recounted. One can only conjecture what could have induced this young man to send his outfit ahead, leaving him to follow along and unprotected, carrying a large sum of money.

Could it have been that his host had provided and he had partaken of too much of the fine liquor that even in such remote places was usually kept on hand in those days, thereby dulling his intellect to the prospect of danger, for surely this young man was lured to danger.

New Mexico at that time was a hot bed of crime and bloodshed. Great cattle barons ruled ranges that embraced territory out of which states might have been made, and their cattle were of uncounted thousands. Cattle rustling was an organized industry, lawlessness reigned supreme. The rifle and the six-shooter talked.

Coming from such surroundings, surely this young man could not have been of an innocent and confiding nature. The very fact that his father had entrusted him with this charge would refute that, and yet, he, his horse and equipment disappeared as if the earth had opened up, swallowed them all and closed again, and to this day it remains one of the unexplained mysteries of these early times.

Kramer and Sons outfit ready for the circle, 1885.

Chapter VII

The Round-Up

The fall, after returning from Colorado, I went to work for John Mueller of Dodge City, a German boot maker. His shop in Dodge City was a busy one, for Muller boots were worn by all the cowboys in the southwest. He had a herd of cattle on a tributary of the Sawlog, north and east of Dodge City and the brand that he gave was M boot, made as follows: ⌐⌐

With a companion I lived in a small dugout and rode line throughout a snowy winter. Once I became quite ill of tonsillitis, in those days it was called quinsy, so my companion offered to go to town to get some medicine for me. Alas, on reaching the bright lights and stimulating refreshment of the city, he forgot all about me and I was left to my fate. For three days and still longer nights I lay there alone, too sick to even keep up the fire in our cheerless little dugout, and I nearly died.

At the end of that time he returned, still under the "influence" bringing with him only a bottle of whiskey. He had entirely forgotten my medicine, but by that time I was getting better and before long was able to be out about my work again.

There were a few settlers even then up in that country and I made the acquaintance of several. One, that I especially remember, was an old German who lived on the north side of the Fort Dodge reservation on the line I rode daily.

He had a pair of buffalo heifers, nice and gentle, that he was breaking to use like an ox team. I do not know where nor how he acquired them, but he must have raised them from the time they were young calves.

Toward spring, on pleasant days, he was always out with the novel "ox" team plowing or breaking sod. All went well until the buffalo became warm and thirsty, then dragging plow and loudly protesting plowman in their wake, they calmly but surely made their way to the nearby creek.

The antics of the irate old German and the mixed language he used on these occasions made a whole show. This was to me one of the diverting incidents in some otherwise pretty dull days, for the life of a line rider was just routine work day after day.

In 1880, a number of ranches were established in the vast grazing district that lay south and southwest of Dodge City. There had been nothing over all this territory, after the buffalo herds had been killed and driven out, so the grass was a marvelous growth, and attracted by the possibilities of this region, ranchers began to appear, though at first these were widely scattered.

The rancher simply picked out the location that he considered best suited to his purpose and established imaginary lines to indicate his range. As the ranches all were located on streams, the "range" extended a certain distance along the stream and on each side north and south of the stream to the middle of the divide that separated it from the next stream. This was the customary range layout.

During the first year or two and while the range was almost limitless, it was easy for these first ranchers to hold their herds, summer and winter, by means of line riding. But when numbers of herds were brought in, so that every man's range adjoined that of some other, it became difficult to keep the cattle from mixing, especially during winter storms. And so in time there arose the necessity for a large general round-up.

The settlements of ranchers along the streams, south of the Arkansas River, in 1880, as I remember them were as follows:

Kramer and Sons came in on the Beaver River and ranged from the mouth of Clear Creek to the mouth of Sharps Creek.

The S-Half-Circle ranged from the mouth of Sharps Creek to the mouth of Palo Duro and on Chiquita Creek. Hardesty Brothers were the owners.

Above them on Hackberry, a tributary of Palo Duro, the Arnold Brothers had a ranch.

Above the Hardesty Brothers on the Palo Duro was the Three Seven Ranch (777).

It was in 1879, that the Cator Brothers, Bob and Jim, located in the head of Palo Duro. They were Englishmen.

On the Beaver River, about the mouth of Palo Duro, the O X outfit now settled and ranged on the Coldwater and San Francisco Creeks, tributaries of the Beaver River, headquarters on the Beaver River.

Charley Grimmer also settled here and ranged with the C X outfit. Soon after becoming established at his ranch he went back East and returned with a bride, the first woman in all that country. It created a lot of excitement in the cow camps and cowboys rode in to the Grimmer home from far and near to meet and welcome the young wife. In the lives of these men, so far removed from home and feminine influences, this was an event.

It was only a short time later that Jim Cator, who had left his betrothed in England, sent for her, met her I think at Dodge City, where they were married and she also came to the distant ranch to make a home. A Post Office was established at the Cator Ranch about this time and named "Zula," Texas.

I have already told of the ranches on the Cimarron River up to 1878, and now in 1880, the following additional ranchers settled on up the river:

From the O'Neil ranch west to what was then called the headwaters of the Cimarron River, was a Texas outfit. A man named Baily was in charge of the cattle.

Frank Spencer's ranch was at the mouth of the North Fork of the Cimarron River.

At Wagon Bed Springs, John O'Louglin settled and ranged up to Sand Wells. His brand was Pig-pen seven, #7.

Beattio Brothers, known as the J.B. Outfit, ranged from Sand Wells west to the Point of Rocks, and beyond for a few miles.

There were other ranches farther up, but I was not familiar with them.

The winter of 1880-81, a great many cattle from the Smoky River, north of the Arkansas River, drifted down onto that stream and instead of turning them back toward the home ranges, these cattle were shoved across the river and allowed to drift on south at will.

Knowing that cowmen had now settled along the Cimarron and Beaver Rivers, the northern ranchers now relied upon gathering these cattle in a general round-up the following spring.

We started about May 1st and went to Adobe Walls in Texas on the South Canadian River for a starting point. Each outfit had a chuck wagon with a chuck box at the end, the lid of which let down to make a worktable for the cook. On this he cut his meat, kneaded his sourdough bread, and other preparation for the meals that were always cooked over a campfire. The wagon carried coffee, sugar, flour, bacon, navy beans, canned corn and tomatoes and other necessities. Besides the supplies the wagon carried the bedding. Every man had his bedroll and "war bag," a bag containing tobacco, extra clothing, if one had any, and such like. So the wagon was always well loaded and sometimes it was necessary to use packhorses to carry a part of the bedding, these when packed were turned loose and driven with the horse herd.

Every man had his string of horses, usually six to eight. Each outfit had a horse herd. When in camp and the horses were grazing, the "wrangler" was supposed to help the cook in the matter of hustling fuel and so on.

When all were assembled at Adobe Walls for the 1881 round-up, it comprised of some 400 men, for besides all the ranchers along the Cimarron, Beaver, Canadian Rivers and intermediate points, some came from as far away as Pueblo, Colorado. There were a hundred men from up on the Smoky and Arkansas Rivers in Kansas, to gather up the cattle that had drifted down from those points during the winter.

At this first big round-up, adjustments had to be made. It was found necessary to appoint a round-up "Boss" whose word would be law for all. The Smoky River men attempted to put in one of their men for this position, but this was over ruled and a man of our own territory was chosen. O.N. Tyson, foreman of the Bullard Outfit, and a real cowman was the one appointed.

Each evening, he outlined the work for the following day and riding from wagon to wagon gave directions for the next day's work. One range was worked each day; a few were so large that it took two days to cover the ground.

Early in the morning, everyone was active. A rope was stretched from the wagon and against this rope barrier, the horse herd was driven in by the "Wrangler," then every man roped the horse of his string selected for the day's work. These were saddled and bridled and left to stand with hanging bridle reins while the men ate a hasty but substantial breakfast. Then all except the cooks, horse wranglers and day herders, started out on what was called "The Circle." This was merely a cordon of men that extended around the entire range, all driving in the cattle to the one designated point where the round-up was to be held. The "Boss" was at this place ready to give orders and oversee the work.

Sometimes so many cattle were brought in it was necessary to divide them and make two round-ups. When the cattle were assembled riders were appointed to hold the herd in check while others rode in to cut out the different brands.

Cows with calves were cut out first, each outfit cutting out the cattle of their own brands and having men to hold the cuts. At times there were disputes as to brands, but these difficulties were always adjusted in some manner. The cattle belonging on the range worked each day were turned loose to roam at will for they were on the home range.

In the meantime, the chuck wagons and the day herds, herds being held as they were gathered for the return to their own ranges, had moved on to the place selected for the following day's work. The horse herds were generally brought to the round-up so that the "cutting" horses could be caught and used for their work, for "cutting" horses were seldom used on the circle.

A "cutting" horse was trained for his work, his rider merely starting him to cut out an animal from the herd. After that the main business of the rider was to keep his seat in the saddle, for these horses were so quick on the turn that they would follow a dodging cow like its shadow, and besides watching for prairie dog and badger holds. At the very best though, someone usually got a fall during this work.

The round-up over for that day, each outfit took their "cuts" and drove them to the scene of the next day's work. Here they would find their wagons and day herds.

The cooks had meals ready for the hungry crowd for there had been nothing to eat since breakfast and the work was always finished before stopping to eat. This meal was served any time from noon on to three o'clock or later. Fresh beef was the main part of the meal.

It was customary for the man whose range was being worked to furnish the beef for the crowd and so two or three calves or a couple of yearling steers were slaughtered daily. This mid-day meal generally consisted of plenty of strong black coffee, great kettles of stewed beef with soup made by adding canned corn and tomatoes to the beef broth, and sour dough bread baked in Dutch ovens. Sour dough bread made by a good cook was excellent. A starter of flour and water was mixed in a stone jar and allowed to stand until it reached a certain stage of ferment. When the day's baking was being prepared, all but a small portion of this was used, adding salt, shortening, soda, water, and enough flour to make a stiff dough which was well kneaded, made out into large biscuits. They were put into the Dutch oven and allowed to stand for a short time to rise, then baked by a slow fire of live coals.

Dried navy beans were considered a prize necessity, and the kettles of beans and bean soup well seasoned with plenty of good bacon was a meal no hungry man could find fault with when cooked well done. And that brings to mind the tale of how the Cimarron River got its name.

As is well known, beans will never cook tender in hard salt water, and the water of the Cimarron River is just that. The tale runs that a couple of old plainsmen camping on the banks of this stream one night placed a pot of beans to cook over a slow camp fire and kept it going all the night through. Next morning every bean rattled hard as hail on their plates and in deep disgust they packed their belongings and pulled out leaving the bean pot and all over the fire. As they drove away one of the grizzled old fellows shook his fist at the offending pot of beans and shouted "Simmer on, d—- you, simmer on." This story may be true, but that was not how the Cimarron River got its name.

However, to get back to the round-up of that spring, there was such an immense scope of country to cover. It soon became evident that the work could best be accomplished by dividing into two parties, one going east from Adobe Walls and the other west from that point.

All the men from the Smoky, Arkansas, Beaver and Cimarron Rivers went over on Wolf Creek in Texas and worked down that stream to its mouth at Camp Supply. Here the men again split, one outfit working up the Beaver River, the other crossing over to the Cimarron River and working that stream, all meeting again at the Point of Rocks far up on the Cimarron River.

It would have been impossible to trail all these cattle over all the country traversed by the round-up crews, so from time to time brands of cattle would be sent to their home ranches. Line riders would then hold them until round-up time was over and all had returned home once more.

Then a home round-up was held, calves were branded and about August first, beef shipments were started and these sometimes ran on at intervals up to November.

The "Mavericks" found on the ranges during round-up time were young animals, generally yearlings, bearing neither brand nor earmark and not following a cow. These were considered the property of the man on whose range they were found. It was his right to mark and brand them in his own brand. A "Slicker" or "Sleeper" was an animal that was earmarked but not branded. Each outfit had an earmark to go with their brand so, of course, these animals belonged to the outfit whose earmark they bore.

A Texas man told me the story of how "Mavericks" got their name. It was during Civil War days in Texas that a Texas cattleman named Maverick found help scarce that many of his calves remained unbranded, neighboring ranchers knowing this when running across these unbranded animals always remarked, "Oh, that is one of old man Mavericks" or more often, "That's a Maverick" and so the term "Maverick" came into use.

Chapter VIII

Mr. Holloway

I worked at the Egg Half Diamond Ranch all of the autumn on 1881. A Mr. Holloway was in charge, a man probably fifty-five or sixty years of age. He was an odd and unusual character but we all liked him. Dave Mackey was working for Holloway also that summer. Dave and I were both young chaps, others on the ranch were all older men.

Just after I reached the ranch and before starting on the spring round-up, Mr. Holloway decided he needed more cattle and wider range. So taking me with him, we made a trip far up the river to where a Texas herd, owned by a Mr. Bailey was being held and cattle and range were being offered for sale.

In reality, the range was the property of no one and was as free to one person as to another, but the man who settled on it first claimed it by prior right. If he wanted to go out of the business, cattle and range were offered for sale, but if he wished merely to move on to some other location with his cattle, the range he had occupied was simply vacated, or left for any one who might come on later and settle on it.

So now Mr. Holloway decided to look this proposition over. We made the trip in a spring wagon carrying bedding and food along and we took also a couple of good saddles and horses. Reaching the Bailey Ranch we stopped over for several days while he looked at the lay out, but he did not buy. However, we did not return to the home ranch until we had a bit of fun and adventure.

At this time, far up on the Cimarron River on this Bailey Ranch and some of the adjoining ones, great numbers of wild horses roamed. They had their ranges on the great flats that stretched for miles to the north and south of the river.

Each band of mares and colts were led by a stallion and had a customary range over which they roamed and certain watering places they used. Some of the bands were small, others large. I saw some that comprised as many as fifty or sixty animals, for some of the more powerful and vicious stallions went out and met the stallions leading other bands and when these met there was a fight to the death while the mares in a huddled bunch looked on. The victor, in these terrible battles on the plains, then acquired the band that had belonged to the vanquished one and so his harem grew.

When danger threatened the band, the stallion rounded up his band and with flying darts to the right and left drove them at top speed. If one lagged a little he rushed in and with wicked ears laid back nipped the laggard with vicious dabs of bared teeth.

On this great plain, north of the river, there roamed a snow-white stallion, one of the most beautiful animals I ever saw. And with another band farther on was one coal black. The fame of these two superb animals was carried far.

Many men looked at these magnificent creatures with envious eyes, but it was almost impossible to catch the stallions, as a usual thing, one of the first things hunters of wild horses did was to shoot the stallion, for when the herd was hard pressed he would stop to fight the pursuers.

Catching wild horses was one of the sports of the times, then and for some following years. But as the country became more settled these animals drew farther away to wilder regions.

Several methods were used to catch them. One of these was by stalking, men working in relays keeping the band ever on the move, giving them no time for rest or food or drink, day or night. Moonlight nights were selected for the work, following relentlessly after the band as they circled around and around their accustomed range, for unless forced to they would not leave it. Sometimes this was kept up for a week, until from sheer weariness, they allowed their pursuers to come close enough so that at last they could be driven into a prepared enclosure and captured. The colts when taken young were soon tamed and gentled.

Another method that required less time, and more skill was "creasing" one horse of the band. This method was used when only one horse of the band was wanted and it was considered the easiest way to acquire a wild horse.

"Creasing" an animal was accomplished by means of a rifle shot that was aimed to make just a superficial flesh wound along the cord that extends along the top of a horse's neck, stunning the animal for a short while, during which time he may be haltered or otherwise secured for handling. Of course, this required great skill and accuracy in shooting, but hunters actually did acquire wild horses in this manner in those days.

Part 2 • Chapter 8

So now, while we were in the wild horse country, Mr. Holloway determined he would capture one, just for the sport of it. And he felt very confident of his ability to "crease" an animal.

Our plans all laid, we repaired to one of the watering places where one of the bands came daily to drink. Mr. Holloway secreted himself nearby, rifle in hand, while I was stationed on horse back just out of sight over a sand hill. The rifle shot was to be my cue to dash in and while the horse lay stunned, securely halter him, and with strong rope fastened to my saddle horn, my well trained horse was to hold and subdue the wild one.

In due time the horses came to drink, led by a fine bay stallion. The rifle shot rang out and the stallion fell. I dashed in and with utmost haste put on the halter and secured the rope, then taking time for a better look I saw that the animal was bleeding profusely, shot through the heart.

I jumped on my horse as if all were well, and in a minute or so Mr. Holloway came up. But when he saw the results of that shot, his face was a study. "Say, boy," he said, "I want you to promise me you won't tell this on me to the other boys, 'tarnation' that was a bad shot!"

"Oh, sure," I said, "I won't tell." Oh no, but wouldn't I though, that was too good to keep. Of course, the story was well spread as soon as we got home and the old man never heard the last of it while he remained in the country. So as a wild horse hunter he failed to qualify, but he was always doing some odd and unexpected thing. One of these was the note he wrote us in the sand.

We were branding calves on the ranch; just gathering up small bunches of cows and calves, taking them into the corral for branding. The branding pen, by the way, was made in a narrow washout back of the ranch dugout, by the simple expedient of perpendicular walls; the front was then closed by rawhide throngs. When the cattle were driven into this place there was little danger of them breaking out, but I must say, it was a very uncomfortable place for us to work, for no breeze could penetrate and we fairly sweltered as we worked in the hole.

One morning, we started out to bring in a bunch for branding, three of us going down the river where most of the cattle would be found. Mr. Holloway making a wider circle was to meet us at a certain point. We gathered our cattle and waited for his appearance but in vain. After a long wait we became uneasy, thinking some accident had befallen him, perhaps his horse had fallen and hurt him. We turned the cattle loose and spread out fan wise riding back to look for him. I felt really uneasy, but Dave, who had been with him longer and knew his ways better, said there was no cause for alarm. Sure enough, on reaching the ranch, there was the old man. He had shot an antelope, carried it home across his saddle, dressed it and was now happily engaged in frying antelope steaks for our dinner.

We were feeling a bit peeved over our long wait and uneasiness about him and told him about it. "Hell," he said, "didn't you find that thar note I writ you down thar in the sand at the mouth of the drawl? I told you I shot an antelope and was coming on home with it and for you fellows to come too."

"How the devil," I asked, indignantly, "did you expect us to find writing in the sand when we were not looking for it?"

"Where are your eyes boy?" he answered. "That writing is down thar in the sand just as plain as can be." And it was too, for we rode by there the next morning just to see it. In large letters traced in the smooth sand with a stick was the old man's "note."

Returning from the round-up on the Canadian River that summer, Mr. Holloway and I reached the Beaver River ahead of the others. He was in a hurry to get to a round-up that was to be on the Arkansas River in a day or so. We had the chuck wagon with us. Camping near the river was a freight outfit waiting to get across for the Beaver River was on a rampage, and crossing with a wagon was impossible. Mr. Holloway told me to wait there with the freighters until such time as I could get across, as for himself, he was going on.

I knew he could not swim a stroke and that stream looked wicked. I protested his going. "You can't swim, you are liable to drown in there," I told him. "Shucks boy," he answered, "I won't drown, if my horse goes down I will just slip off and crawl out on the bottom."

I can see him yet, getting ready for the plunge in. He took some paper money out of his trousers pocket and folding the bills carefully placed them in the sweatband of his old hat, then tying boots and trousers onto his saddle, he mounted "Old Warrior" his favorite horse and plunged into the seething waters.

"Warrior" was equal to the task and they reached the opposite bank in safety. Waving me a gay goodbye, the old man was off on his long ride to the Arkansas River.

I camped with the freighters that night and we all found it possible to ford the river the next morning. When we reached the Carey and Lane Road Ranch, on the south bank of the river, we were told that a Mexican had been shot and killed near a spring on the Arroya, north of the Beaver River, along which the Jones and Plummer Trail led.

We reached the place and camped there before noon. A path down a steep incline led to the spring and going down this path hurriedly for a pail of water, I came upon the body of the Mexican so suddenly that I could not stop and leaped over it. The stench was sickening for the body had been lying there about a week.

The manner of his death as we had heard it was as follows: Two freighters with wagons loaded at Dodge City for Tuscosa were joined by this Mexican who was on his way to the same destination. One of the wagons was loaded with barrels of whiskey. After getting out on the trail one of the barrels was topped and the three men becoming intoxicated got into a quarrel. The Mexican was shot and left lying where he fell.

Calling to my companions, I reported my find and all came to view the gruesome remains. One of the men had a load of bones on his wagon and he now picked up the head of the Mexican, with his pocket knife extracted the bullet from it and then tossed the head up on his load of bones with the remark that "every little helps". But his companion with more decency forced him to put it back and we then gave the remains a shallow burial nearby. And that was how Mexican Arroya, north of Beaver City, got its name.

The men who were responsible for the killing came back a few days later and had by then become somewhat alarmed as to the possible consequences. They then took up the body and taking it to a spot in the sand hills, a mile or so east of the spring, they once more buried it.

Years later, when the country had been settled by homesteaders, and the town of Beaver City was quite a thriving little place, the local paper came out with a story of a skeleton found in the sand hills north of the river. The sand had blown away, partly exposing the bones and someone had found them. I knew at once that it must be the skeleton of the unfortunate Mexican. At this late day when all the principals in the episode have passed on I deem it no harm to tell the story.

Day Herd on the Cimarron. R.K. Perry in the foreground.

Chapter IX

The GG Ranch

When calf branding was over that year, we made an early beef shipment of eighty head of three and four year old steers. Taking about three days for the drive to Dodge City, we reached there only to find the Arkansas River at flood stage. It was arranged to cross the steers on the bridge, an unheard of thing in those days, but Mr. Holloway was noted for doing unheard of things.

Hiring three or four extra men to help, we bunched them up close to the bridge. Then I was sent in ahead and the steers shoved by the other riders followed me.

When the steers heard the hollow noise, made by many hoofs on the wooden bridge, they became panicky and started to run pell-mell. I had to keep out of the way, reaching the north side I turned the fleeing herd east along the bottoms in the direction of the stockyards. A wild run followed that led us through back yards and around dwellings and business places.

Many a person on foot raced madly to get out of the way. When we finally got them penned at the stockyards we found we could not ship out that night but would have to wait until enough cattle came in to make a trainload. It was not until the next night that the train was ready.

In the meantime, our spring wagon was loaded with supplies for the ranch in readiness for an early start for home the next morning. When the old man left with the cattle that night bound for Kansas City, he instructed me to get back to the ranch the next day "if I had to kill a horse to do it." His orders were followed to the letter.

At sunrise next morning I was on the Mulberry, about twelve miles south of Dodge City, driving the team on the spring wagon and leading three saddle horses. At noon I stopped for an hour and a half for rest and food for man and beast and again was on my way, reaching the ranch about ten o'clock that night, having traversed a distance of some sixty-five or seventy miles.

The horses did not show any sign of great fatigue for they were the hardy Texas ponies of that day and their endurance was wonderful. Never the less, when we got out the next morning, Dave found "Old Ronie," one of the driving horses, down near the river, dead.

Mr. Holloway, when told of this on his return to the ranch a few days later, said not one word. I had obeyed his orders and his rush in getting me back to the ranch was that we were to round up the adjoining ranges and get out our cattle for a second beef shipment. Mr. Holloway had learned on reaching town that beef prices were up and he was anxious to get the benefit of the higher rate. So very soon we were on our way once more, this time with a herd of two hundred and fifty fat cows and big steers. On reaching the Arkansas River at Dodge City, we found that stream was still high, but shoved the cattle in and prepared to follow.

I was riding a sorrel horse we called "Dopay," a beautiful and intelligent animal. Someone had trained him so that when his rider would extend his hand and snap his fingers, the horse would rear up on his hind legs and walk.

When the beef herd started swimming water in midstream they began to mill; around and around they circled, eyes bulging, horns clashing, bawling and spouting water from their nostrils. I happened to be on the up-stream side of them and now dropped into the "mill" with the cattle. How dangerous this was, I only realized after getting into the mass of closely packed animals, horns had to be dodged and I had to take my feet out of the stirrups at times to keep from being creased, had my horse gone down it would have been "goodbye." Twice we circled before I saw a chance to pull my horse out and head him for the north bank and by some miracle the frenzied cattle followed him out. I had taken a long chance and won, but I vowed "never again for me."

When all were safely across, the old man rode up and said, "You darn fool kid, why did you ride into that mill like that, you might have been killed."

"Well," I answered, "That's the way to break up a mill of cattle in high water. I took a chance and won, didn't I?"

"Yes," he said, "you won, but by Gee, if you had lost, you would be shoveling coal in hell right now."

Mr. Deardoff, the owner of the Egg Half Diamond cattle died at Kansas City that summer and the estate was being settled. So now cattle and range were offered for sale and on Mr. Holloway's return from Kansas City, after this last beef shipment, he brought with him a prospective buyer, a Mr. Hays, who on looking the proposition over bought the outfit and took charge.

Part 2 • Chapter 9

The following Spring, he bought more cattle and extended the range from the Cimarron River to Crooked Creek, about four miles south of the present site of Meade, Kansas. He established headquarters ranch there, re-branding all the cattle, choosing for the new brand the crooked ell. $\mathcal{L}$ A year or two later Mr. N.E. Steele was placed in charge and the Crooked Ell was known as the Cimarron and Crooked Creek Cattle Company.

When Mr. Holloway turned the outfit over to Mr. Hays, he offered jobs to all the old hands but I had already been offered a better position at the Fred Taintor Ranch. About September 15, 1881, I began work there and for the next eleven years the G. G. Ranch was the place I called home.

Taintor Creek, as it soon came to be known, was a small spring fed stream of pure clear water. A tributary of the Cimarron River, it flowed almost directly north sinking away in the sand before reaching that stream.

About two miles of the creek, which lay along high bluffs, was heavily wooded with immense old Cottonwood trees. Blackberry, China berry and Willow trees also grew there and there was thick undergrowth in places. Elderberry, plum thickets and wild grape vines grew in abundance.

When I first went there it was a wild and lonely spot. In former years it had been a favorite camping ground for bands of roving Indians out on hunting trips. The brakes and underbrush were full of deer.

The ranch house was made part dug out and part logs. It was located near the mouth of the timber and on the west bank of the creek. It had a fireplace at the back and was lighted by a small-framed opening for a window that as yet had no glass.

There was also a small stable made of poles with a thatched roof. The corral was also made of poles and these were lashed together by means of rawhide throngs.

Mr. Taintor, then a young man of about twenty-two, was a native of Hartford, Connecticut. On his graduation from Yale, he had come West and engaged in the cattle business. First in the Comanche Pool, selling out there and coming here the summer of 1881, he bought the Bullard herd and range. Fitzgerald Brothers who had ranged with Bullard were, at the time I reached the ranch, getting out their cattle to be moved farther up the river to the ♀ (o cross ell) owned by W.I. Harwood, who had bought the Bailey Ranch that summer. Harwood and the Fitzgeralds all were Bostonians.

At the same time, Mr. Taintor's men were getting out a bunch of fat cows to ship. On leaving with them I was left in company with an old Texas trail hand to ride line and hold the remaining cattle on the home range.

There was a pet goat on the ranch and he proved a great pest for each night when we were sleeping soundly he came in, by way of the small window, and helped himself to our sourdough bread and anything else cooked or uncooked that he could get at. Nosing about amid pots and pans he would waken us and if any one arose to put him out he invariably leaped lightly to a large box and from thence through the window opening.

I, determined to break him of this habit, fixed up a trap so I could close the window by means of a board on rawhide hinges.

The second night after the outfit had gone from the ranch I caught Mr. Goat and pulling the board down over the window I proceeded to give him a well-deserved beating. The goat, in his efforts to get away, jumped into the lower bunk where my companion was sleeping, trampling him ruthlessly with his sharp hoofs. This so enraged my co-worker that the next morning he saddled his horse and pulled out for Dodge City, leaving me alone with 1500 head of cattle to keep within bounds. He was tired of this "blankety blank" country anyhow, he said, and was going back to Texas.

And so for thirty days I was left alone and I mean alone, for not a soul came near the place.

The outfit was detained in town as witnesses on a lawsuit. The summer's work was nearly over anyhow, so there was no rush about getting back and town life was very pleasant for a change.

Each morning I got up early and after a hearty breakfast, for there was no noon meal, I started out on my fifty-mile line circle. Going south up the creek I turned west across the flat to the western range limit. I then headed north across the Cimarron River, going still north across Crooked Creek and to the top of the hills beyond it, then east to the eastern range limit, from there south to the range limit, then west again to my starting point at the head of the home creek, turning in all cattle found on my way that were heading out from range bounds.

Coming back down the creek, I always drove up the loose horses, caught and picketed a fresh horse for the next days work. By the time I had cooked and eaten my supper I was ready to call it a day and was able to sleep soundly the night through.

In all the thirty days of my lone stay at the ranch I saw but one human being. Riding down into the Crook Creek valley one day I saw a man with team and wagon getting cottonwood poles from a little wooded side canyon. He told me that he and several other men had been sent in there to make a dugout for George Anderson, a cattleman, who was preparing to move in at that point on the creek.

As we were talking, I noticed his six-shooter in its holster and saw that it was cocked. I said, "Mister, your gun is cocked, it's liable to go off." "Oh," he answered carelessly, "I carry it like that all the time." I road away in disgust.

When my outfit got back from Dodge City, and all the late calves were branded, we turned the cattle loose for the winter. Work was over for the season. We loafed, hunted, trapped and visited at neighboring ranches far and near. Deer were plentiful in the timber and canyons, bands of antelope ranged on the nearby flats. Whenever we wanted fresh meat we went out and got it. Also, we had fresh mutton whenever we cared for it as a little band of sheep, perhaps twenty-five head, came in on the creek and wintered there close by the ranch house.

These sheep and many other small bunches that were scattered all over the country at that time were strays that had fallen out of herds being driven through the foregoing summer, for during 1881, vast numbers of sheep were brought up from western Texas and New Mexico to Dodge City. Herd after herd passed through here, some were shipped to eastern markets, some sold to local buyers at Dodge City and many were driven farther north.

An enterprising person might, that fall, have gathered up a sizeable flock of these strays, if one cared to do so. Sheep were scorned by the cowboys except in the form of muttonchops, so these strays in time all fell prey to the coyotes.

Fred Taintor

The Taintor Boys

Part 2 • Chapter 9

GG Ranch stable and corral built in 1882.

GG headquarters on Taintor Creek, built in 1882. The hospitality of this old ranch was known far and wide.

Chapter X

The Locoweed

The winter of 1881-82, was wild and open, there was no loss in cattle. Grass was plentiful and well cured and there were no bad storms to drift the cattle far from home ranges. Green grass came early that year so that we started on the general round-up by April 20th.

As in the previous years, men were sent down from the Smoky River to our roundup but not in such numbers and as the cattle had not drifted far from home. Our work was much lighter that year.

The spring of 1882 was memorable however, for it was the time that locoweed made its appearance in great abundance. It sprang up as if by magic. Its soft gray green leaves and purple clusters of blossoms made an innocent picture and little did any one suspect its true nature, nor the inclinable harm it was destined to inflict on cattle and horses alike.

Once an animal got started eating of the loco it soon became an "addict." No longer would it eat grass but walked from bunch to bunch of the weed, eating greedily. Very soon its effect became apparent, the body of the animal became emaciated, the head enlarged, the eyes had a glassy stare and the beast walked jerkily lifting the feet high. It was only a matter of time until death ensued.

The herds coming up from Texas that year also suffered from this pest, for the cowboys ignorant of its effect, seeing great patches of the stuff thought it a species of clover and would leave their horse herd to graze where it grew rankest.

"This is such fine clovah you all have up heah," one of the Texas boys remarked to me one day. And would scarcely believe the "Clovah" was poisonous. That spring on the round-up the Indians put us "wise" to the evils of loco. They called it "Heap Pony Suffer."

The Indians with their traditional lore, handed down from generation to generation, were able to tell us many things about this country that we were ignorant of. They claimed that every seventh year was a bad loco year.

The effect of the weed was so widespread and disastrous that the cattlemen sent for veterinarians to come out and examine some of the victims. It was found that the brain and all the organs were affected but an analysis of the weed at that time failed to disclose the poison in it. For a time it was believed that a small worm that wove a fine web over many of the plants was what caused the trouble. But in later years, with better facilities for analysis, the poison was found in the plant itself.

That spring Mr. Taintor purchased these different brands of cattle, the Bar Links, M C's and Oil Can brands. These were bought from a Mr. Farris, whose range was on Bear Creek, Clark County, Kansas. We gathered these and brought them home and re-branded them in the **GG** Brand.

Martin Scully and his wife came to work on the ranch that year. Mrs. Scully cooked for us and was the first woman on a ranch on this part of the Cimarron River.

After the general round-up Scully and myself rode line to hold the cattle until calf branding and beef shipments were over. Coming in one evening from my line ride with rifle across the saddle in front of me, for I meant to get a deer to take home that evening, I rounded the point of a small draw and came suddenly face to face with an Indian. We were both startled and instantly I had my gun trained on him. "Me scout," he hastened to explain. "Soldiers down on creek," and he pointed in that direction.

"All right," I said, "show me." And he rode ahead while I followed and coming to the top of a bluff overlooking the creek I saw a company of soldiers, Negro Calvary from Camp Supply encamped below us. I rode down to the camp and found that it had been reported that Indians were leaving their reservations and coming north so this company had been sent out to find and bring them back. They had with them several of Uncle Sam's Indian Scouts and it was one of these I had encountered.

The soldiers remained in camp next day while the scouts were sent out reconnoitering but no Indians were found this far north. The officers, of course, were white men and the second evening at our invitation, they came to the ranch for supper and to spend the evening.

As they were leaving in the morning, I rode by their camp as I was starting on my line ride. My horse, an unruly beast, excited by the unusual sight started pitching and bucking in the camp. The Negroes delighted with the "exhibition" laughed uproariously and yelled, "Ride him."

Breakfast was being prepared for the men over a large campfire, a forked pole at each end held an iron rod, from which was suspended large iron kettles containing coffee and food. Into this my plunging steed now made his way, upsetting the whole mess into the fire, where upon the bugler ran into his tent, snatched up his bugle and gave the "breakfast call." It was all fun for the black boys, and when I got my horse quieted and rode up to the officers' tent they too were laughing. Their breakfast had been served so they did not share in the campfire calamity.

The Martin Scully family left us that fall for they had two children, the oldest having reached school age. They decided to move someplace where schools were available, so we lost our woman cook and for a number of years thereafter had men cooks at the ranch.

The Loco was so bad that fall that many of the cattlemen fearing to turn their horses loose on the range, as in previous years, took them elsewhere to be wintered. So it came about that I was sent up in the Crooked Creek valley settlement with the Taintor horses. Here there was very little of the locoweed. The grazing was in fine condition and there was hay also. I turned the horses out to graze during the day, but every evening fed them hay. Two other men were with me. Howard Wright, in charge of the Crooked Creek horses, and a man from the K Ranch on Wolf Creek in Texas. Bull Joe, we called him because he was holding a bunch of bulls up there for his outfit. We rented the home of Labe Windor, a bachelor homesteader, for the winter. He turned his place over to us and went back East for a visit.

We were a gay lot, there was not much work to do so there was much time for play. I fear some of our pranks rather shocked some of our conservative neighbors.

Our place soon became a stopping point for all the cowboys to the south on their way to and from Dodge City, so there was always something doing, some fun going on. We made the acquaintance of all the settlers, attended the Crooked Creek church, a lady preacher was in charge of it that winter and that added interest to our attendance. "Bull Joe" took an active part in the meetings, he was a good singer and knew all the old gospel hymns and how he would sing. On occasion he even got up and made "talks" to the congregation.

A preacher from an adjoining neighborhood sometimes came to the meetings, his mode of transportation was a homemade sled, a box with a hinged lid was fastened on the front of the sled for a seat. One night we placed a number of empty whiskey bottles in this box where they remained for sometime. When finally discovered by one of the good sisters of the church the preacher had a hard time accounting for this damaging evidence.

A short distance from our "soddy" was the home of a Norwegian bachelor who could speak but a little English. Everyday he would come to see us, all though we were constantly playing tricks on him. On several occasions we went to his place at night, after he was asleep, and piled sod up before his door, clear to the top and moved up every bulky object we could find besides, so that in the morning he had several hours work before he could dig his way out, but all was taken good naturedly.

It was that winter Frank Biggars made the ride that was famed in cowboy annuals.

Joe Morgan, a cattleman with a ranch on the Canadian River, had been to Dodge City and while there went with the officers to arrest a man living some forty-five miles south of the town. On reaching the place, a dugout, and seeing no sign of life about, they entered and found the man dead of Small Pox.

On returning to his ranch home, Joe Morgan was soon stricken with Small Pox in a most virulent form. His wife was frantic, the nearest doctor was at Dodge City, one hundred and eighty-five miles away. So Frank Biggars, foreman of the "K" Ranch, a neighbor of the Morgan's offered to make the trip to Dodge City for the doctor.

Leaving the Morgan Ranch he rode to the "K" Ranch where he changed horses, going from there to the Kramer Ranch on the Beaver River where he got a fresh horse that he rode to our place on Crooked Creek.

While the boys got Frank something to eat I caught and saddled one of our best horses for him. And this one he rode in to Dodge City, making the entire ride of 185 miles in about sixteen hours.

At each stop for a change of horses he had made arrangements to have a team in readiness for the return trip with the doctor. Accordingly next morning we took a team to the Hoo-doo Brown Road Ranch, on the Jones and Plummer Trail at the Crooked Creek crossing.

Stopping at Dodge City only long enough for the necessary arrangement for a team and doctor, Frank was back on his way south. But it was a losing race for when about halfway on the return trip they were met by another rider to inform them that Joe Morgan was dead.

Chapter XI

Outside Man

On returning to the ranch that Spring I was sent to the round-up as "out-side man." Out-side man for an outfit was a man who was sent much farther out than the wagon would go, sometimes such a man would go as far south as Red River, east to Anadarko, west to Pueblo, Colorado, or north beyond the Arkansas River. Not every outfit sent an out-side man, but when one went he was expected to bring back with him any cattle found belonging in his part of the country, no matter by whom owned. So that year I was out the entire summer, returning to the home ranch only once or twice and then just to stay overnight.

I took with me seven horses, my bed roll and "war bag" carried by one of them, and proceeded to Cantonment, Indian Territory, about sixty miles east of Camp Supply. I traveled down there with the "D Cross" outfit and wagon.

Arriving at Cantonment, some days before the round-up was due to start, we camped while outfit after outfit came rolling in from Comanche Pool, Medicine Lodge, Anthony, Caldwell, Cherokee and Ponca Counties and all over this part of the cattle country and the Arkansas.

While waiting here the time was spent in hunting; deer and wild hogs were plentiful. Everyday there were games and contests of various kinds; horse racing, foot races, wrestling, roping, and riding contests, and, of course, for those who did not enjoy the strenuous sports there was the ever present card games and that popular old pastime "mumbledty-peg."

There were always numbers of Indians encamped close by Cantonment and as they were fond of sports they too took part and were especially good at horse and foot racing.

We never wholly trusted the wily Indians for all their friendliness, and always night herded our horses down there. For the Indians would, on occasions, stampede and scatter them, thereby getting a chance to slip out a few that they might keep and use for a time at least.

One night there was a severe storm that I well remember, for I was on night duty. Several hours of stormy wind and rain, during which the horses drifted with the storm and the herders, in the darkness, could only drift with them. So herd after herd met and mingled and there was a grand mix-up of hundreds of horses.

Next morning we had a horse round up and each outfit cut out their own horses to be held separately, as was the custom.

I worked from Cantonment to the head of the Beaver River that year, meeting my own outfit at Camp Supply. Working with them as far as Sharp's Creek where I again left Coldwater Creek and down this stream to its mouth and then up the Beaver River to the Anchor D Ranch. From there across to the Cimarron River, a long drive of about sixty miles across a high flat, that was un-watered except when heavy rain falls left ponds and lakes over this vast stretch of fine buffalo grass grazing land. At such times the cattle from both the Cimarron and Beaver Rivers would wander far out and range on this fresh grazing until the ponds were all dried up and they were compelled to seek the river country for water.

At the time we crossed this sixty-mile stretch, between the two streams, I was with the Patterson wagon. They were from Sand Wells, about halfway between the H.D. Ranch and Wagon Bed Springs on the Cimarron River.

We had a herd of about a thousand steers, cows and calves. Leaving the Beaver River one afternoon about two o'clock we filled up everything we could find to carry water in but had no water bags. There was only enough water to last us for supper and breakfast. At noon we stopped and rested, had a lunch of sour dough bread and bacon, but there was no coffee or water. Driving on that afternoon we saw many gleaming, glistening lakes in the far distance, seeming to invite rest and refreshment. However we knew it was only the elusive mirage that can be so tantalizing to the thirsty traveler, and it soon became evident to us that the lakes and water holes had all been dried up by the heat and drought.

Mr. Patterson was an old trail boss and under his strict supervision the herd was carefully handled, not pushed, but eased along with as little fuss and confusion as possible. It was entirely due to his excellent management that we made that long drive without water, in the heat of mid summer without the loss of so much as a young calf.

That night we made dry camp, the men were too thirsty to eat much. The cattle were bedded down and I was on second night shift with a fellow we called "Six Shooter," when there came up one of the worst electrical storms I have ever experienced. The huge black cloud

Part 2 • Chapter 11

extended from horizon to horizon, thunder rolled and roared and crashed, lightning played in long snaky zigzags and again in broad blinding flashes. Everything seemed electrically charged, for balls of fire glinted on the horns of the cattle, static on the tips of our horse's ears. I stroked my horse's mane and a line of fire followed my hand.

Six Shooter rode around to me and in an awe struck voice said, "Say, what is this, every time I spit it looks like a stream of fire, and just look at the herd, fire playing all over them." Neither of us had ever seen the like before and we were just plain scared, but there was just one thing to do and that was to keep on riding around and around the herd to keep them as quiet as possible and so the storm passed on.

We had hoped for a good rain but only a few large drops came splashing down, soon the stars were shining and all was serene once more.

As soon as it was light enough to see next morning we were on our way. No one wanted any breakfast so the cook was instructed to drive ahead to the Cimarron River, make camp and cook a good dinner for the men. We figured we must be within fifteen miles of the river and the cattle, suffering from thirst plodded wearily on, not caring to graze.

When within three miles of the river a breeze sprung up from the north and the cattle catching the scent of water threw up their heads and the big steers in the lead ran full speed for the river. Many of the cows, too, left their weary calves and followed. The men all racing in with the cattle leaving me at the tail end with a few cows and a mass of nearly exhausted young calves to keep pushing along.

Men, horses and cattle all plunged into the river and drank together. When finally, their thirst satisfied, two of the boys came straggling back to where I was still working doggedly along with my charges and asked me what I was doing back there, I was mad clear through. Those boys got a piece of my mind for I was in just the right mood to get quite a load off my chest. When I did get to the river I lay down flat and drank and drank. I never knew Cimarron water could taste so good.

After a good dinner every one felt better. We had reached the river at "Point of Rocks" and that afternoon we moved the cattle to their home range at Sand Wells and left them. Going back to the wagon, that had been left behind, we worked next day about the Point of Rocks and the following day below the Point, then on down to the Patterson Ranch at Sand Wells, where we laid over for a day or two of rest.

There was a family at the ranch, father, mother and a lovely young daughter. As it happened a couple of girl friends were visiting her, so the second night of our stay Mr. Patterson gave a dance for our benefit. To be sure there were only the four ladies as partners for the twenty-five or thirty men, but these four danced tirelessly and almost constantly. When occasionally they did stop to rest there would be a stag dance where half of the dancers had silk or bandana kerchiefs tied on an arm to designate them as "ladies." At midnight there was a fine "feed" that had been prepared by the lady of the house, assisted by the wagon cook. All in all it was quite an event.

When we again started out, our work took us down to Wagon Bed Springs and from there to Frank Spencer's, at the mouth of the North Fork on the Cimarron River. Here we broke up and went home, for another round-up crew had worked up to this point that spring. So bidding goodbye to our friends of that part of the range country, four of us "outside men," with our horses and about twenty head of cattle belonging to different ranchers in our part of the country now pulled for home.

It so happened at this time that none of the cattle we were bringing home belonged to the outfits we represented, for it was the custom for the outside men to gather and bring in all cattle belonging anywhere in his part of the country. And woe to the cowboy who was not true to his trust. For that was a part of the "Code" of the cow country. While we were not bringing any of our own cattle, at this particular time, perhaps at some other distant point other men were working our cattle in toward home for us.

The cattle industry at that time was like one huge family, all working together, all looking out for each other's interests. We felt honor bound to do so.

On reaching the home ranch I found instructions to go on at once to a round-up on Wolf Creek in Texas.

Waiting to go with me was Smith Ruble from the George Anderson Ranch on Crooked Creek. So with a fresh string of horses we started after dinner, and rode leisurely to the Beaver River where we stopped overnight at a line camp of the Healy Brothers.

Doing line riding from this camp were a couple of young fellows who had not been long in the country and were new to all its ways. We called them the Healy tenderfeet.

In the talk after supper they told of seeing bear tracks on a creek south of the river and seemed quite excited about it. Knowing that bears had never been found in this part of the country, Smith and I merely exchanged amused glances, while they talked at some length on the subject. In the morning we started on our long ride and about noon stopped at the head of a small creek where there was a water hole and a lone tree.

Part 2 • Chapter 11

We seldom carried with us a lunch, but on this occasion had done so for we knew there would be no ranch camp or habitation of any kind on our way and it would be late before we could reach our destination. So at the line camp that morning we had wrapped some biscuits and bacon in a newspaper we had carried with us. Having disposed of our food, we prepared to rest for a short time and Smith amused himself by reading the paper that was then months old.

Our loose horses were grazing not far away while our saddled ones with reigns dragging stood nearby. A gusty wind was blowing and I warned Smith to be careful with the paper and had scarcely done so when a gust of wind snatched it from him and with a smack it blew against the legs of one of the saddle horses. With a snort of fright he ran pitching and with stirrups clattering into the midst of the loose horses and away they all went in a wild stampede and were soon out of sight, but headed for home.

We looked at each other in dismay, this was really serious. We had come about twenty-five miles that morning and to think of having to walk back. It was awful. But there was no other way, so after a few not well chosen words, we started on our painful way, stamping along in our high heeled boots that were never meant for walking.

After walking for some distance I saw afar off something gleaming. Sunshine reflecting from some bright object and toward this we now directed our footsteps. About three miles farther on we came upon my horse, standing in a little depression and it was the sun gleaming from my saddle horn that had caught my eye.

The horse was easily caught and I said, "Now, Smith, I will go on and find your horse." "No you don't," said he, "not from here, but we will go back to the place we started from so you will have some land marks whereby to find me when you get back." So taking turns about riding the horse, we returned to the water hole and the lone shade tree. Here there still remained a biscuit from our noonday lunch and stowing this carefully away against possible hunger, Smith admonished me to look carefully about, as I rode away, and fix the spot in my memory so that I would not fail to find him, in that trackless scope of country when I came back.

Then dispersing his length on the ground in the shade of the tree he prepared for a nice undisturbed nap, for Smith was a chap who could take it easy. On cool mornings when out on the round-up he always put on an overcoat and wore it till noon, even though the sun came out blazing hot. He always contended that what would keep out cold would keep out heat, but the boys said he was simply too lazy to remove the coat.

I left him lying there and galloped off in pursuit of the horses. Ten miles farther on I found Smith's horse, standing with a dejected air, his mouth cut and bleeding. In trying to keep up with the other horses he had frequently stepped on his long trailing bridle reins, every time this happened it brought him up with a jerk on the bit that hurt cruelly so he had given up and was quite willing to be caught. Returning to the place I had left Smith, I found him still reclining.

The loose horses we knew would not stop until they reached the Beaver River so now we made our way back there and found them not far from the camp we started from that morning. One of my horses was missing. He, I found out later, had gone all the way home.

It was now late evening. We would have to spend another night with the two boys at the line camp. "They will be surprised to see us back, what shall we tell them?" I asked Smith. "Well," he answered, "you tell them we ran onto a big bear and you roped him, but he was too much for you, started to walk up on your rope, so you had to turn him loose and that my horse stampeded. Tell them a real bear story." "No," I said, "You tell the story, and I will back it up. You are a quiet sort of chap and they will be more apt to believe it if you tell it."

So we drove the horses up into the corral and caught fresh ones to ride next day, picketed them out for the night. The others we hobbled and turned out.

Then going into the house, Smith told our story with collaboration and detail furnished by myself. It went over big. "There now," one of them exclaimed triumphantly, "we been telling Healy and the other fellows about seeing them bear tracks, and they would not believe us, now since you fellows saw the bear and roped him I guess they will think we know what we were talking about."

We spent a pleasant evening and after a good nights rest were off the next morning and this time reached our destination without trouble.

The tenderfeet left no time in telling the bear story to all they met but whenever it was told the listeners always enquired. "Who were these fellows that roped the bear?" And, on being told invariably broke into loud and delirious laughter so that it soon became clear to our hosts that they had been duped. They were fighting mad about it, and that was another mistake they made, for with the cowboys the fellow that could not take a joke was out of luck, to show anger only added to the delight of the tormentors. So now the bear story was spread far and wide over the ranges and those boys were teased most unmercifully.

Smith and I worked down Wolf Creek and then up the Beaver River. When the round-up reached the Healy Range we were met by the tenderfeet. They were still so mad that they wanted to fight, but it takes two angry parties to make a quarrel and we only laughed at them, telling them they should not believe everything they were told.

But there is a sequel to this story, well, perhaps not really a sequel, but the fact is that a year or two later we really did have an encounter with a bear. A real live one, too.

Far away on the Beaver River, where bears were never known to have been, but how pleased these boys would have been, what unalloyed joy would have been theirs, could they but have witnessed our ignominious flight up that blind canyon, hotly pursed by a harassed little brown bear. And, last but not least, they would have found deep satisfaction in the rich quips and jokes that assailed us from all sides, as, our work over, we all gathered around the chuck wagon and were served with bear steak for dinner. But unfortunately our young friends missed all of this that would have been as soothing to their injured pride for: "He laughs best who laughs last".

We however, were able to enjoy the fun and jokes at our expense for to us it was all in the day's work and play.

As for bear meat, in its time and place, it may have had its good points. But for a bunch of western cowpunchers accustomed to the best beef on earth, those bear steaks proved a bit disappointing. They were not the rich treat we had been led to expect from our reading of the prowess of famous bear hunters of old and their subsequent feastings.

But, I must be getting back to my original subject that I left away back yonder in order to tell a bear story or two, for I was telling of my many long trips as "outside man" on the round-up that year.

After leaving the Healy Ranch, we then worked up the Beaver River to the mouth of Sharp's Creek and there turned back home. I only stayed at the home ranch a day or two and then was off again, this time going to the Gorham Ranch on the Cimarron River, about twenty-five miles southeast of the present site of Protection, Kansas. From there, the round-up worked back west, working Bluff, Kiowa, Calvary Creeks in Comanche County, Kansas.

At the mouth of Bluff Creek we split, one party worked up the Cimarron River, the party I went with worked up to the head of Bluff Creek in Clark County. We brought back the cattle we had gathered over the old "Mount Jesus Trail" to where Ashland is now and then worked Beverly Brothers range and D.S. John's range on what is now John's Creek going from there back to the Cimarron River where we met the party that had worked up that stream. We went on up the river working range after range until we reached the ♀ Ranch about fifty miles up the river from our home ranch.

This ended the round-up season for that year. There only remained the late calf branding and gathering the beef herd from the home range to drive to the railroad at Dodge City for shipment to western markets.

ZH outfit, Southeastern Colorado, taken in 1884.

Day herd leaving the Beaver at the mouth of Sharps Creek, No Man's Land.

Chapter XII

Back on the Ranch

I have told, in some detail, of the country I road over that summer. To the casual reader this might not seem so great an area. But the man on horseback, riding for days, weeks and months, seldom retracing any of his way, revealed the immensity of this range country.

Only a short time before, the home of vast herds of buffalo, disturbed only by occasional bands of Plains Indians on hunting expeditions, it retained a form of virgin loveliness almost untouched by the hand of man.

There were a few scattered ranches along the streams, scarcely making a break in the vast solitude. There were no east and west roads. By day we traveled in the general direction in which our destination lay. By night the North Star was our compass and guide.

One might ride whole days without sight of another human being. Over great plains, merging into rough hills that led to wooded small streams feeding the large rivers. All was beautiful, unspoiled nature, with such an abundance of rich grass lands as cannot be imagined at this day.

It was beautiful alike, in the bright green of early spring, the darker shade of midsummer and the soft brownish gray of autumn. Sleeping under the stars for a month at a time, we learned the positions of the principal stars and grounds of the firmament.

Standing night guard we sometimes saw strange and awe inspiring sights, falling stars, meteorites shooting across the heavens in a blaze of light and, on rare occasions, an exquisite moonbow.

Have you ever seen a moonbow? It is like a rainbow only being reflected by the softer light of the moon the colors are more delicate. A moonbow is a thing of real beauty.

It was a fascinating life, living in the open with the beauty and appeal of nature almost undisturbed by human kind. I truly believe that it was these conditions that developed some of the richest personalities as well as some of the most unique characters I have ever met.

During my absence that summer a new ranch house was built, a two room, cement building. Captain French of the Crooked Creek settlement with a couple of other men from the same place as helpers put up the house and to this, a year or so later, another room was added. This room was made of sandstone laid up in mortar and plastered with cement, making in all a long low well-built house that still stands.

A mower and hay rake were brought down from Dodge City and I think I am safe in saying that these were the first implements of that kind to be used near the Cimarron River.

Three or four stacks of wild hay were put up, down on the river meadow. As we had no wire for fencing, a rough protection was made of poles lashed together by rawhide so that in winter we would have hay for our horses.

Winter was usually playtime on the ranches but that winter we worked at getting out posts and fencing a horse pasture. In the spring we also fenced a hay meadow on the river. A freighter was kept busy all the winter bringing out wire, grain, and supplies for the ranch. Each trip he made he brought the mail from Dodge City not only for our own place but also for some of the adjoining ranches, and thereby hangs a tale. One of us discovered the matrimonial publication, "Heart and Hand" and this discovery proved a "Bonanza" of amusement for several months.

"Heart and Hand" was entirely given over to the "Art" of bringing together those who were matrimonially inclined although a few of the appeals for correspondents specified that the object was "fun" with "perhaps" a view to matrimony if the right one could be found.

One dollar was the accompaniment for each name sent in with a short description, sometimes a photograph of the applicant. A reading of "Heart and Hand" was quite diverting.

Unknown to the other fellows several of us got together and sent in the names of some of our friends to the publication asking for correspondents "with a view to matrimony." The men we chose for this prank represented two types, the illiterate fellow who would be pleased and flattered by the attention. A man just the exact opposite and the latter type included the bachelor owner of a neighboring ranch who came regularly to our place for the mail each time the freight wagon was due.

The response we got from our appeal to "Heart and Hand" was gratifying to say the least. The first mail brought a mere handful of letters for each of the men whose names had been sent in, but when the second mail arrived there was a full gunny sack of letters from the fairer sex.

Part 2 • Chapter 12

At the first mail these letters were read by the recipients and stowed away in pockets without comment while we looked on apparently without interest. But when the bag full arrived it was too good to keep and all of us shared in the fun. We, who were responsible, kept up such a show of innocence that we were not even suspected, and it was not until much later that the truth of the matter became generally known.

Those long winter evenings were passed in smoking, talking, storytelling, and song. Not much reading, for the light from the fireplace and the tallow dip was all the illumination we had. Always there was someone in camp who was a good singer and some of the popular songs of that time as I remember them were: Lone Star Trail, Sam Bass, Home on the Range, Bury Me Not on the Lone Prairie, The Dying Ranger, and others that I have now forgotten.

A good storyteller was always a welcome addition to the fireplace group, such a one was "Happy Jack." He was just what his name implied and we never knew him by any other or where he came from. Though he must have been around a lot for he knew many places and had a fund of stories that was never exhausted. He would not work, though he always had a job during round-up time, but he just went along and shirked everything that was disagreeable. He always got out of standing night guard and when left on day herd invariably deserted and would be found loafing around the wagon.

He never bought tobacco or cigarettes, yet always had plenty of both. He never owned a slicker, but if a rain came up Jack always kept dry and some other fellow could not find his slicker anywhere.

King's Jester for the cow-camps during the long winter months, he drifted from place to place happy and carefree, borrowing a dollar here and there, never paying back any but he was never entirely out of funds. Once he left us and went to El Paso, but soon he was back. Said he did not like it there for they made him work. He had a whole set of new experiences and now told them with gusto.

One of these was that soon after he reached El Paso he was "broke," as he had no acquaintances from whom he could ask a "loan," he wrote on a card: "Please help an unfortunate - I am deaf and dumb." Armed with this he repaired to the residence section, stopping at a neat little cottage he knocked. When the man of the house come to the door he handed him the card, the wife too now appeared and looking over her husbands shoulder read the message.

"Poor fellow," she said, "deaf and dumb."

"Yes, Ma'am," said Jack, eagerly, and then remembering his role, not stopping for his card, he turned and ran. The "deaf and dumb' racket was off.

So, at the first opportunity he returned to this country where his "talents" were better appreciated.

His happy impudence always got him the best of everything without apparent effort until one day some of the boys riding up to a plum thicket found Jack filling up on the luscious ripe fruit while some distance away his horse stood waiting with bridle reins hanging. This was a chance too good to miss. One of the boys rode up and grasping the reins of Jack's horse led him off leaving Jack to walk the three miles to camp. Now this was a real blow but there seemed no help for it so he started on his weary way. The boys were in high glee. At last Happy Jack had to do something that was a real effort.

While still some distance from camp, Jack saw someone coming in his direction leading his horse. Good fellow. Someone had felt sorry for him and was bringing him his horse. But as he drew nearer the good Samaritan apparently did not see Jack who waved and shouted to no avail, something was wrong with both the sight and hearing of his "kind" friend and so after all, Jack was forced to trudge all the way in to camp on foot. But, that was the only time anyone got the best of him.

When he finally left us for good, no one ever knew where he went, he just drifted away and we never heard of him again.

It was before moving into the new ranch house and while still living in the old dugout that the boys at the ranch pulled off a stunt that was perhaps fairly typical of the times.

We had a visitor from Dodge City, a young man new to the west and desiring to see something of ranch life. The first morning after his arrival he asked where he might find the blacking brush. I explained that we did not have such an article on the place but that we sometimes greased our boots with tallow.

This innocent question, however, served to convince us at once that our visitor was very green and as time went on this impression was emphasized.

He talked a great deal about Indians, perhaps to his eyes the surrounding seemed fitting for skulking redskins. At any rate we began to think he should not be disappointed in having an encounter with Indians and so it was arranged.

As it happened we had several visitors that day who came riding in from distant directions. All of the cowboys from adjoining ranches and each of these were met out at the stable and given pointers on conversation so that all reported having sighted Indians during their ride to the ranch.

Of course, this caused much excitement and discussion so that by evening our guest was all worked up. "Well," he said when bedtime

Part 2 • Chapter 12

came, "what are we going to do about the Indians?" "Do?" someone answered. "Why, we will put up a fight if they show up here."

It was early autumn and we were still sleeping outside, some of our beds were out at the stable. One man had his bed roll further away under an old wagon, and as everything had been arranged he now took the young fellow out there to sleep with him.

Scarcely had they settled themselves in their bed when a shot rang out. "What's that?" asked the guest, in a hushed whisper. "Indians," was the answer. "And, we had better get to the dugout." Jumping up in great excitement the poor fellow raked his back on a projecting bolt on the bottom of the wagon bed.

By this time a perfect fusillade of gunfire was going on, accompanied by a few blood-curdling yells.

I was standing by the corner of the corral as our man came by on his way to the house. Afraid to stand erect, lest he would make too good a target, he was loping along on all fours and grunting, "Ugh, Ugh, Ugh," reaching the dugout still on all fours he tumbled in.

A sick man (he was really sick but he too was in on the secret) was sleeping inside. "Hey, what's the matter, what's all the racket about?" he asked. "Indians," was the gasping answer, "I saw dozens of them as I came up from the wagon."

After a time the shooting died down and we all came straggling into the dugout and told how we had put the Indians to rout. After some discussion it was decided that we should go out and station ourselves at different points to guard against further attack. Thrusting an empty gun into the hands of our visitor he was left to guard the house. We, of course, went off peacefully to our beds, but the sick man reported to us next morning that our man walked the floor all night long, carrying his empty gun.

It so happened that our wagon was going into town the next day and our visitor, who had intended to make a more protracted stay, now climbed up beside the driver and announced that he would be going back to town.

A year later I was riding into town and met the young man out for a walk. He had a position then in the Santa Fe Offices. I stopped and we talked for a while. Finally he said, "Say, I know now what you fellows pulled on me. I figured it all out afterwards when I got on to the ways of the country and it's all right, you had a lot of fun. What hurt me the worst was that six inches of skin I raked off on the wagon bolt."

So after all, he was man enough to take his "initiation" without rancor or ill will.

Bar 7L, Colorado outfit taken in 1884.

Chapter XIII

On The Roundup

In the spring of 1884, a number of new ranches appeared here and there along the streams. Now the ranges were not quite so unlimited, as they had been, yet there was plenty of room. Some of the older ranchers sold out and moved away, new ones took their places and still the herds from Texas came up in great numbers. These herds being driven through to railroad points in Kansas, furnished some interesting incidents of those times.

I was riding range one day and when near the Tuttle Trail I was mystified by hearing the sound of many bells. Bells were not unusual at that time for we all used them to some extent. One horse in the bunch was usually belled at night and the freighters put bells on their oxen or mules when turned out for the night, but this was a sound of many bells and it was daytime.

Following the sound, I soon came upon a herd of eight hundred head of Texas cattle, fully one hundred head of them wore bells. Talking with the boys, I found that those cattle were from the rough brush country in Texas and the ones wearing bells were "rouges."

These cattle, raised in the brush and running wild the year around, had become so attached to their habitat that it was almost impossible to drive them away from their home. Some of these steers were ten to twelve years old, huge rough long horned fellows, they had been rounded up many times and placed in herds to be driven to market. But they had developed the sly trick of slipping quietly out during the night and returning to their favorite haunts, so now several of the ranchers determined to get rid of these old timers, rounded them up, belled everyone of them and put them in the beef herd bound for Dodge City. There was no more slipping away at night, for when a bell tinkled away from the herd a vigilant night guard was soon on the trail, so the scheme worked very well and now these big old long horns were nearing the end of their last trail.

In a previous chapter I have mentioned O.N. Tyson as being round-up boss of our first big round-up. Mr. Tyson was an all around cowman well liked by everyone. He was a handsome, well built man and walked with a slight limp due to the fact that he wore an artificial limb. This was not generally known until the following incident occurred. He was in charge of a beef herd that was being driven from somewhere in Texas to the railroad at Caldwell, Kansas.

This herd had developed the habit of stampeding day or night at the slightest provocation or even none at all. It was a hard trip for all hands and Mr. Tyson finally tried out the plan of scattering the herd along a stream at night time instead of bedding them down and close herding as was the custom when on the trail. Worn out from long hours in the saddle, Mr. Tyson rode into the wagon one evening when the herd was at the mouth of Bluff Creek in Comanche County, Kansas. Picketing out his horse close to the wagon he lay down and was soon sound asleep. In a short time the Negro cook awakened him, "Mist Tyson, de cattle runnin' agin."

Jumping up only half awake Tyson ran to his horse and forgetting to slip the picket rope noose over the horse's head he sprang into the saddle and applied the spurs. In some way the picket rope got looped around that artificial leg and when the horse reached the end of the rope there was a jolt that brought him up short and wrenched the leg from its fastenings. The leg went hurtling through the air, landing near the campfire where the cook was at work.

With a terrified shout the Negro ran to hide under the wagon. "My Lord, My Lord," he wailed, "Mist Tyson done lost his laig." Tyson now free of the rope, with his one good foot in the stirrup raced on to head off the stampeding herd.

Of course, the incident created a lot of fun. "The leg" sometimes developed a creaking noise in some of its joints and then its owner would sing gaily this song of his own make-up: "Tyson had a wooden leg, so they say, Tyson had a bacon rind laid away, just to grease that wooden leg, so they say."

He was one of a number of men I knew in those early days that did not allow a physical defect to prove a handicap. He was only more determined to make good and to be a leader in his chosen work.

In the spring of 1884, the round-up started again at the Dickey Ranch a few miles above Cantonment. We were on the grounds early, waiting for all to assemble. Near us was the encampment of Chief "Stone Calf" and his band. Unlike the majority of the Indian race Stone Calf had a genial disposition, he liked to talk and entertain. He had been to Washington, D.C. at the time treaties were made with the tribal

Part 2 • Chapter 13

chiefs and he now displayed, with great pride, some copies of these treaties he had brought back with him. With a magnificent gesture to indicate distance he always referred to Washington, D.C. as "Washington, way up in Kansas."

In friendly rivalry, many of the cowboys engaged in horse racing and gambling with the young Indians and that night Stone Calf held a tribal dance for our entertainment. But the next morning a few miles farther down the river, close to Cantonment a more sinister drama was enacted.

A herd of some seven hundred head of horses, coming up from Texas, to some northern point had stopped here for the night. For some time previous to this, it had been customary for the Indians to demand a tribute of a horse or cow from the owners of herds to pay for the privilege of passing through the Indian Nation. This had always been granted for it was much the easiest way to avoid trouble. Chief Buffalo and his band were near and when Chief Buffalo made the usual demand of a horse from this herd the owner refused. An argument followed, the Chief stating that he would take one anyhow. Where upon the white man warned him "If you ride into that herd to cut out a horse, I will kill you."

Without regarding this warning the Chief rode into the herd and cut out the horse of his choice, then the white man, true to his word shot and killed him.

This happened about ten o'clock in the morning. A number of Indian were camping in the vicinity, excitement ran high. Serious trouble between white men and red men seemed imminent.

The men with the horse herd now realizing this, abandoned the herd and took refuge in a stone-house nearby, dug out portholes and being well armed, prepared to defend themselves.

Word of the trouble was brought to the round-up encampment by a couple of anxious cattle owners of those parts. The wagons were drawn up close together and half the men were left to guard them and the horses. The other half were sent into Cantonment to help out there in case of trouble, as only a few soldiers were stationed there since it was merely a place for distributing rations to the Indians.

There was a telegraph line by then from Fort Supply to Fort Reno and a message was sent to Reno asking for troops. A company of cavalry started immediately for Cantonment.

In the meantime envoys were busy with both sides. The Chief of Scouts going about among the Indians and quieting them as much as possible for they were gathering in sullen vengeful groups, while some of the cattlemen went to the stone-house and talked with the men there.

In this way hostilities were averted, but that night the situation was tense, no one slept, had a gun been accidentally discharged the effect could have been instantaneous.

At sunrise next morning the troops arrived, hungry and short tempered after their all night ride. One of the officers now carrying a white flag went to the stone-house and placed the men there under arrest. Stopping only for breakfast and to feed their horses, the Calvary men started the return trip to Reno with their prisoners.

Knowing that with the arrival of troops, the horse herd as well as the men with it, would be taken in charge. The Indians, with fine forethought, stole the entire herd, breaking it up into bands of about a hundred and driving them off to distant points.

We did not hear what punishment, if any, was given to the man who killed the Chief, but we did learn a year or two later that the Government paid for the horses stolen by their Indian wards.

The R-S outfit on the Beaver River, near Camp Supply, had a government contract to supply beef for the Indians at the issuing post at Anadarko. At stated intervals, a herd was to be delivered to that place. Going through the Indian country each trip, the Indians exacted the tribute of a beef, and this in time became irksome.

Working for the outfit at that time was a young fellow called "Robin Hood," what his real name was I never knew. But, in general characteristics, he was much on the order of our "Happy Jack." Somewhat lazy and shiftless, he was as well a clown. Used as a sort of roustabout, he was generally left to look after the ranch during the outfit's frequent trips with the beef cattle. Hearing the complaints of asking tribute payments, he asked to be taken along on the next trip, promising that he would "put an end to that foolishness." Accordingly he was taken on their next trip and he made good his promise.

It was well known that Indians feared a crazy person and would always give them a wide berth, so now Robin Hood using his talent for clowning played crazy. Whenever an Indian or a group of them were sighted he left the herd and rode to meet them. When close up he put on his crazy act, leaping from his horse and performing a series of antics, somersaults, hand springs, contortions and so on. Leaping back on his horse again and going through more crazy actions, while his onlookers gazed in amazement, ending by making a run at them. The Indians invariably fled and were soon far out of sight. His unique scheme worked to perfection and was highly diverting for his associates as well.

To return to our round-up of that year, we followed the usual routine. This season was marked by many severe storms, always

Part 2 • Chapter 13

occurring at night. Rain fell in torrents and the thunder and lightning that accompanied it was far out of the ordinary. Night herding was dangerous work, stampedes and accidents of horses falling with their riders were of common occurrence. At the mouth of Crooked Creek, where it empties into the Cimarron River, a young man was killed by lightning during one of these terrific storms.

The autumn that followed this summer was a particularly pleasant and peaceful one. We started out early for the fall beef round-up and reaching the meeting place east of Camp Supply, where Woodward now is, we camped for perhaps ten days waiting for the others to arrive. We spent the time in fishing and hunting. Large channel cat abounded in the Beaver River and on the little creeks emptying into it were perch and catfish in abundance.

Wild turkeys were plentiful and very fat at that season of early September. Sometimes we went to their roosting place at night and shot down a few, but we liked best to come upon a flock in daytime and ride them down. They would run along ahead of us for sometime, when they began to tire we ran up on them, some too fat and tired to attempt flight could be caught. If all flew we had only to follow after them, for a wild turkey when fat will make but one short flight. After that they are easily caught, so our wait was very pleasantly spent.

When all had arrived and we started work, following the usual route working the Beaver, Cimarron and Crooked Creek Ranges, branding late calves and gathering the beef cattle, finishing our work on the Crooked Creek range. Part of our men were sent home with the day herd while we, with the beef herd and chuck wagon, started for Dodge City.

Camping the first night near Hoo-doo Brown's place, we held the cattle on the little flat where Meade now stands. The herd had made a run on us that day and when night came were still excited and nervous, milling and moving about constantly. We boys put in rather a hectic time of it, and sang the night through to quiet our restless charges.

Returning from beef shipment, we made ready for another winter. It was about this time that Mr. Taintor went East and bought fifty head of fine registered Hereford cows and a couple of herd bulls that cost him $500.00 each and shipped them to the ranch. We kept them in the fenced horse pasture that winter and gave them extra care.

At that time when the cattle of the country were almost exclusively Texas long horns, these registered animals were an innovation and provoked a good deal of laughter and comment. But the wisdom of this pioneer movement for better grade cattle became apparent before many years had passed. Mr. Taintor was a pioneer not only in this business of better grade cattle but also in several others that made for the betterment of the country. He was one of the first to build up and improve the ranch home from the crude dugout to substantial and comfortable buildings of rock and cement and that included barns and corrals as well.

From conversations with old time plainsmen and hunters we know that our little timbered creek had long been a favorite haunt of Indians and white men, but in working about the place we found evidence of a former occupation of the place that antedated any of these. Almost completely covered by earth and grassed over, was what appeared to have been a large stone building or corral and farther away a sunken spot, evidently a dugout, with just a rock fireplace still showing. At different places in the country, there was evidence that at one time, before the advent of the Indian and the buffalo, some pre-historic race of intelligent people had inhabited these vast plains.

About three miles from where Englewood, Kansas now stands, there could, at that time, be plainly traced the outlines of an ancient irrigating ditch that had been taken out at some spot on the Cimarron River and led off to the northeast over a fine buffalo grassed flat.

It was evident too, that in some past time the bed of that river had lain to the north of the sand hills that lay east and south of Englewood. On Wolf Creek, in Texas, there were the remains of a small adobe town. Neither Indians nor plainsmen could furnish any explanation of these things.

Sometime in the past, while the buffalo still roamed here in great numbers, Pete Lamblin, an old plainsman and buffalo hunter, in the course of a hunting trip came to what is now Taintor Creek. Coming over the crest of a hill, he saw below him on the creek an Indian encampment. Something out of the ordinary was going on and the Indians, when they sighted him, waved him back. This was a warning that in those days was never disregarded, so he retired from sight. But being curious to know what was going on, he left his wagon and creeping to a good vantage point, lay flat and watched while the red man buried one, evidently a Chief, with most impressive rites. His curiosity satisfied, Pete made his way back to the wagon and fearing discovery lost no time in putting the place behind him.

Now, after a lapse of years, he came back to this part of the country. Having become acquainted with Mr. Taintor, while he was ranching in the Comanche Pool country, Pete came to the ranch to make us a visit. Incidentally, too, he had kept in mind the burial of the Chief and his treasures and now meant, if possible, to find the spot and have a look at what it contained.

Part 2 • Chapter 13

We, of course, were interested in his story and now turned out to help find the Indian's burial place. A half-mile or so above the ranch house, on the high sloping hillside of the east creek bank, just outside the timber that lined the creek bed, we found a grave. Nearby was a crystal clear little spring of pure cold water.

Sure that we had found what we were searching for, we used spades and shovels and soon uncovered what lay within. But this was no Indian grave. In it was a rough box and when we pried off the cover off it, there, in a fair state of preservation, was the form of a white woman and around her forehead was bound a man's large red kerchief.

On seeing this, Pete knew at once whose body that lay here. He had heard of this woman's tragic death at the time it happened. She was the wife of one of the old time buffalo hunters. With her husband and two grown sons, she had come out on the hunting trip that in time led them to this spot, an ideal one for camping. In this lovely sheltered place, well watered and timbered, an abundance of game all about, one would have thought this family might have spent a restful interval here. But no, one morning the boys became involved in a bitter quarrel, when one of them snatched up a hatchet to attack his brother the frantic mother flung herself between them just in time to receive the crushing blow to her own forehead.

The three men cut up their wagon bed to make a coffin for the mother. So with the handkerchief bound about her head to hide the cruel wound, they laid her in her hillside grave and left her to her long lonely sleep.

No loved one ever returned to lay a flower on the lowly mound, or stand for a time in silent tribute. But the sad winds of autumn whispered a requiem through the tall tree tops. Winter spread a snowy protecting mantle over the spot. Spring brought her offerings of rich perfume of blossoming plum thickets and the fainter, sweeter odor of wild grape blooms and that of the delicate pink wild roses that grew nearby. The fierce sun of mid-summer was tempered here by the cooling shade of the great cottonwood trees and so the cycle of the seasons passed.

Long she had lain there in her narrow bed, undisturbed by the shy wild creatures that passed it by. The doe and her fawn came from their leafy covert and picked their dainty way to the nearby spring. Feathered songsters built their nests and caroled in the tree tops. And now this peace was violated by curious seeking men.

With respectful care we replaced the boards and mounded up the earth and there she rests to this day, but the grave of the Indian Chieftain has never been found, though at times search is still being made for it. With Indian cunning, the place had been so disguised that the white man was never able to relocate it.

On the round-up at Chiquita Creek, No Man's Land, Oklahoma, in 1885.

Chuck wagon and cowboys at dinner on Chiquita Creek, Oklahoma,

Chapter XIV

The Blizzard of '86

The spring round-up of 1885 was a very large one, for many new ranches had settled wherever an opening could be found, for the cattle industry at that time was reaching its peak. Many herds had been brought up from Texas the previous season and sold at Dodge City, turned loose in the fall to roam at will and by spring were badly scattered.

Again it was a season of storms and much rainfall, on the uplands every buffalo wallow was full of water and in places large lakes were formed. The new settlers coming in at that time would not believe that these lakes were not permanent. When told the contrary insisted that it was merely a ruse to discourage settlement.

It was a busy season for the cowboys, we had to work the country over several times that summer in order to get the calves all branded and beef cattle gathered up. That fall it seemed everyone was eager to own more cattle, even the boys working on the ranches invested their savings and bought a hundred or two cows or yearlings, branded them and turned them loose just as everyone else was doing.

Up on the high flat country south of Dodge City there was now plenty of water and a number of herds were held there around the lakes until time to turn them loose in the fall.

New settlers, too, had come in all over the country, especially in the Crooked Creek Valley and in Clark County. That summer and fall a drift fence was built, running east and west on the line between Texas and the Neutral Strip, starting at the east end of the strip where it joined the fence, it extended west to the New Mexico line.

The purpose of this fence was to keep the cattle from this country from drifting into the south Canadian country and so simplify the round-up business. The cattlemen from this country had been forced to send men and wagons into the south Canadian country each year to gather up the cattle that had drifted so far from the home ranges and this drift fence seemed a solution for this problem.

However, the fence proved the "Waterloo" of the cattle business. The fall was a very pleasant one, long and lingering, the days perfect, week in and week out. But it was becoming evident to us all that the ranges were over stocked, the fine abundant grazing was no longer what it had been. Still, everyone was feeling happy and optimistic for cattle prices had been going up and the winters had, in general, been fine and open. But this year many of the cattle brought up from Texas had been in very poor flesh and the yearlings were known as "Dogies" or "Dogy" for they all looked stunted and undernourished. However, they all went to "Dogy" heaven that winter.

The 6th day of January 1886, Beck Boland, who was working for the Ranch came to the GG Ranch and proposed that he and I take a trip around the country. We would visit at various ranches and see how cattle and grass conditions were, for it was becoming more and more evident what the grass was getting short. No feed had been put up except a little hay for the saddle horses.

It was a wonderfully fine day that 6th of January. Beck and I started out after dinner and rode to the Healy Ranch to stay that night and we rode over there in our shirt sleeves as it was too warm for coats.

The three Healy brothers, Frank, George and Nat, were all at the ranch when we got there. Their ranch home was a large dugout with a board floor and a fireplace. We spent a sociable evening smoking and talking and about ten o'clock went to bed. After retiring to bed we heard the wind come up and knew it had turned to the north, but soon fell asleep. About 3 a.m. I was awakened by the intense cold. Everyone else was awake too. Finding it was impossible to sleep because of the cold, I dressed and started the fire. But my best efforts with the fire did not seem to improve the temperature of the dugout very much. Everyone got up early and when daylight came the snow was coming down in a thick white mass of fine stinging particles that one absolutely could not face.

Frank got out somehow and fed the horses in the nearby dugout barn. With advancing day the storm increased in intensity, the rising wind blowing and swirling the snow into every crack and crevice. Try as we might we could not keep the dugout comfortably warm. A thermometer at the ranch registered 28 degrees below zero that evening. Just before nightfall there was an occasional lull in the wind's intensity and it was during one of these that we managed to slip out and again feed the horses, but they had to go un-watered. Sometime during that night the clouds cleared away and the next morning the sun came up on a cold white world.

Part 2 • Chapter 14

Beck and I had intended making about a hundred mile trip that would have taken in a number of ranches and time was no consideration, but our plans were all changed now. About 10 o'clock we set our faces homeward and a cold trip it was too. By keeping to the ridges where the wind had blown the snow off we got along very well until we struck the high flat between the Beaver and Cimarron Rivers. Here the snow was so deep all over that it made the going hard for our horses but they floundered through so that we reached the J Ranch in the evening. Next morning Beck went on home and I, with two of our men, started south for the drift fence as we expected to find trouble there. It was about twenty-eight miles down there but it took us all day, we had to pick our way along the ridges as much as possible to avoid the deep drifts.

Stopping at a nearby ranch that night, the next morning we went on to the junction of the drift fence with that of the Y pasture fence and here a pitiful sight met our eyes. It was still intensely cold and here in this corner cattle had drifted in by the hundreds. Held by the strong four-wire fence they had huddled and died until living cattle were walking across the fence over the piled up bodies of the dead. All up the drift fence were cattle, both living and dead. We worked the day through driving back the cattle where we could but it seemed a hopeless task so we cut the fence in several places and let the poor moaning, bawling creatures pass through on their hopeless tramp to the south.

We stayed that night again at the ranch and next morning I left the two men there to do what they could along the fence while I returned home to see how things were going there. On reaching home the boys told me that the country between Crooked Creek and the Cimarron River, north of the ranch, was just one milling, bawling mass of cattle suffering from cold hunger and thirst. The streams were all frozen over, so the next morning early we started out with team, wagon, spades and shovels and began the task of sanding the slick ice to make a crossing for the cattle.

We estimated that we came across 25,000 head of cattle there that day. But no matter how many we crowded on the ice, at no time did it break through. It seemed frozen to the bottom.

The next day we started out in pairs, one carrying an ax, the other a spade and rode into the Crooked Creek country. With the ax we chopped holes through the ice to make watering places and the spade was used to dig out animals stuck in snowdrifts. Everywhere we found dead cattle, hundreds of them. And not only the ones poor in flesh and weak had succumbed, but many large three and four year old steers. Many of them died standing on their feet. It was a strange sight to come upon an animal, after the drifts had melted down, standing with feet firmly frozen to the ground. From a distance they looked like living animals, but they had died thus in the storm several weeks before.

It was twenty days before there was grazing for the cattle, in the meantime they starved or the ones that did pull through managed to barely exist on tops of sage and tall but not nourishing grass tops.

The men I left at the drift fence worked there for ten or twelve days before returning home. Down there the cattle died by the thousands.

The drift fence was 185 miles in length. I am safe in saying that 200 dead cattle to the mile along this fence was a conservative estimate. Some put it higher than that, so the drift fence besides being an expensive experiment proved a complete failure.

Always in the fall a number of men at every ranch were "laid off," that is taken off the payroll, but were welcome to stay on until the spring work again was on. These men now turned their attention to skinning, getting permission from the owners to do so. They went to work on the animals that lay dead everywhere.

Going in pairs and working together they soon became experts at the job and could skin from 20 to 40 animals a day. Letting the hides lay for a few days to dry, they were then gathered up in wagons and hauled to market. A good hide bringing two dollars or more, these men reaped quite a rich harvest. But thousands of the cattle were never skinned and not until the next spring, at round-up time, was the full extent of the loss known.

It was a hard winter for us all. The men who were working on the ranches rode every day all day long, working with the cattle, pulling them out of bog holes, and "tailing up" the weak ones that were "down." Almost invariably when we lent them this kindly assistance they repaid us by getting "on the prod" and chasing us to our horses. But with all our efforts we did little real good for in the end most of them died anyhow. While there were no more blizzards that year there was much snowfall that covered the grass and the extreme cold all during January and much of February weakened them all.

Of course, the greatest loss was in the Texas cattle for they were not acclimated. As an instance of this, I might mention, the case of a brother of D.S. Johns, who had a ranch on Johns Creek in Clark County.

This man lived in Colorado, but came down that fall and purchased 2800 head of Texas cattle and turned them loose on his brother's range. The next spring he came down here, bought an outfit of horses, wagon, etc. and hired men to work the D.S. John's outfit and gather his cattle. Of the 2800 head, they gathered just 38 head. Many a man lost all he had. The loss was almost unbelievable.

That was the end of buying "through" cattle in this country. These southern cattle after living through a winter here became acclimated and were then classed as natives, but after this experience no one wanted to make another attempt.

There were also a number of cattle brought in from Missouri that fall. These did not fare much better than the ones from Texas, for they had been accustomed to feed and shelter and here they had neither.

Round-up on the Cimarron near the Mackey Ranch.

Roundup day at Beaver City.

Chapter XV

On the Way to New Range

That spring, knowing the cattle would be widely scattered and far out, we worked accordingly. Joe King went to work the Canadian and Wolf Creek. Mr. Taintor, Smith Ruble, Johnny Campbell, Zack Fleming, Al Harrington and Charley Noble went with the wagon to work the Beaver Country.

I took George Anderson's wagon and a bunch of his men and worked the Cimarron, Bluff Creek, Kiowa and Cavalry Creeks, the Beverly Range, John's Creek and the Sand Creek country, bringing all the cattle we gathered to the mouth of Crooked Creek where we met the general round-up crew. Then working up the Cimarron River as far as the ⦵ (O cross ell) Ranch, just north of where Liberal now stands.

Vash Musset was round-up boss on the general round-up and Sim Holstein was boss of the Beaver round-up. This was the last year that we worked the Bluff, Cavalry and Kiowa country, for most of the ranchers there now fenced their ranges as settlers were coming in numbers.

Our wagons were sent home from the range, but several of the men were left to cross over to Sharp's Creek to work there. We now helped Ed Budley of the Anchor D Ranch to gather their cattle and take them across to the Beaver River, as these cattle had been sold and were being taken to a new range. There were about 1200 head of them.

The first night we camped about where Liberal now is, and got an early start for our long drive across. The day was intensely hot and we could not drive the cattle too hard so it must have been about ten o'clock that night before we got in on Sharp's Creek and a storm was coming up.

Everyone was tired and hungry for we had had only some crackers for lunch. Now there was no prospect for any supper, although we knew there was a road ranch somewhere in the vicinity, but the night was very dark so there was no hope of finding it. So the cattle were turned loose and it was every fellow for himself and bed down where he would.

Johnny Glenn and I started out together. In the darkness and the rain we soon became separated, so I unsaddled my horse, turned him loose and wrapping up in the blanket and using my saddle for a pillow, weary beyond words I soon fell asleep.

Sometime later I awakened, the rain was coming down in torments, the lightning was terrific. Soon I heard a roaring noise and sitting up I saw by a lightning flash that it was indeed water and that I had been sleeping almost in the creek bed. Snatching up saddle and blanket I climbed to the top of the bank and a little farther on. Once more I bedded down and slept soundly despite the downpour that lasted most of the night.

At daybreak I awoke and was getting to my feet when to my astonishment to see, not more than forty or fifty feet away, a white covered wagon. Walking up to the back of it I peered within. A young man lay there asleep. At my shout of "Hey there," he jumped as if shot and without a look in my direction he dived out head long through the front opening and fell out over the wagon tongue. He was a land seeker and I had "scared him out of a year's growth."

My horse was gone and his mules were nowhere in sight so we were both stranded. Before long though I sighted one of the boys out looking for the horses and they were soon rounded up. The boys appearing from several not far distant points, we all mounted and started in search of something to eat. Down the creek a couple of miles we came to where a couple of "through herds" had made camp the night before. In the storm the cattle had become mixed and they were now separating them.

At the wagons the cooks got us a breakfast of sourdough bread, coffee, bacon and molasses, nothing ever tasted any better. Feeling much refreshed we then gathered up Mr. Budley's herd of cattle and that day were met by the round-up that had worked up the Beaver River. The Budley wagon was along and now took charge of the cattle. We found our own wagon, but not for long were we permitted to enjoy "home" cooking. For at this point all of the wagons from our part of the range turned back home while Sam Kyger, Johnny Glenn, John Sand, myself and several others, each of us representing different outfits, were sent on to work the Beaver, Palo Duro and Coldwater countries and while doing so we ate at the wagons of the ranchers in that part of the country.

The days of the buffalo were over, but at this time, away up here on the Beaver River and its tributaries, there could still be found small and scattered bunches of these animals ranging undisturbed with the cattle. It was while working there that summer that we came upon "Buffalo Jones" of Garden City, Kansas. He was out on one of his trips to secure young buffalo calves for his ranch at Garden City. He was trying out the experiment of crossing the buffalo with domestic cattle, Brahmas, I believe. He called this cross breed "Cattelo."

Mr. Jones had hired a cowboy, Lee Howard, to rope the calves. His method was to bring out with him on these trips several gentle old milk cows to mother the calves when caught. A spring wagon carried their camp outfit and a sort of rack or cage to hold the calves when first captured.

This appealed to us as something entirely new in the way of sport, so one afternoon after the days round-up was over, several of us went out with Lee Howard and Mr. Jones. Over on Wolf Creek we found a small bunch of buffalo and with them were five or six calves. Each of us picked a calf, cut it away from its mother's side and roped it. Some of the horses were afraid of the buffalo and proved rather unruly, others did not show any sign of fear. Mr. Jones followed up with the spring wagon and the little captives were placed in it and taken to camp. When good and hungry the cows were brought in and in a few days the shaggy little creatures with their funny looking "humps" became quite reconciled to their foster mother.

The afternoon's sport made a welcome break in the monotony of the daily work. However, it was our custom when out for weeks at a time to have an occasional diversion, usually in the form of a horse race, and it was later that season that we had a race we all remembered for many a day. It was down on Buffalo Creek, where Buffalo, Oklahoma now is. Joe King, always an enthusiast when horse racing was concerned, had arranged the match between a horse owned by Hi Kollar's outfit and that of one owned by the D-Cross.

We were all backing the Kollar horse, a little roan with a white stripe in his face. He was a great pet and always hung around the wagon where he soon learned to eat everything on the day's menu, but he was particularly fond of sour dough biscuits and so had acquired the name of "Biscuit."

Quite a bit of excitement had been worked up over this race and we had wagered every bit of money we had with us and not only that but all of our trinkets besides, pocket knives, watches, tobacco and so on. We were so sure our horse could win.

The track was laid out, all arrangements made and the race was on. Biscuit took the lead and kept it easily, and then it happened.

John Delmar, the cook, drove the wagon up to a good vantage point, where upon Biscuit promptly "flew the track" and ran to the wagon for some of his favorite provender. Such luck! We were cleaned out, broke absolutely flat and all because Biscuit's love of good food outweighed his racing instinct.

In vain we begged the victors to match another race or lend us a few dollars as "working capital." They were enjoying our predicament too much to lend us any assistance. Biscuit certainly let us down hard.

Not all our afternoons off from work were spent in horse racing and sports, for at times it became necessary to have a "wash day" especially after a hot forenoon's work cutting out cattle from which we all got dusty and sweaty.

After a belated dinner and a short rest a number of us would go to the nearby creek or river where a bath was in order. Stripped of all of our clothing, but to avoid sunburn we were clad in boots, slicker and Stetson hat. The soiled clothing was well soused in the stream, scrubbed with sand instead of soap.

When fairly clean they were spread on the grass in the hot sun where they soon dried and our washday troubles were over and forgotten.

But on one occasion that I well remember, we all got into the ranch from the round-up with a bunch of "cooties" in our clothes. Some fellow had loaded us all up with those busy little creatures so now we must have a real washday to get rid of them. Jack Bateman offered to do all the washing while the rest of us went about some other work we had to do. When we got in that evening and saw our clothes, Ye Gads! Jack had put everything in the wash boiler together; jeans, shirts, socks of various colors, wool underwear in varying shades of scarlet, tan, gray, yellow and even black. After a brisk boiling in strong suds, even Jack himself must have been startled out of his habitual calm by the looks of the conglomerate mess he lifted out of that boiler. Everything had faded and the colors ran into everything else in streaks and globs and splotches.

The boiled wool underwear had shrunk to fit persons several sizes smaller than the owners and instead of being soft and wooly they were now hard as bricks. Still there was a bright side to this incident for we were well rid of the "cooties."

Chapter XVI

The Chuckwagon Chefs

I must not fail to make mention of some of the good round-up chuck wagon cooks of those days. The names of some of them I cannot now recall, but there were Wes Hull, Frank Dole, John Delmar, "T Pete," Bill Anderson, "Jor" Bartlett, and Gus Roberts. Every one of these men was fine in their line.

In order to be popular with the men, the wagon cook had not only to be a good cook but he had to have other attributes as well. He needed patience, plenty of it, a sense of humor, and the ability to carry on under the most trying circumstances of weather and fuel conditions. For instance when one of the long drizzling three day rains came on and all the fuel got wet so it was hardly possible to keep a fire going, or in a gale of wind when a helper sometimes had to rig up a wind break and hold it in place while the cook with smoke filled eyes got such a meal as he could for his hungry crew.

With so many men and horses about, there were sometimes accidents to try the temper of the cook. I remember one evening when we were in camp and had a couple of posts set with a rope stretched between, against which we drove the horses to hobble, some of them for the night. One of the saddle horses got the horn of his saddle caught under this rope and then becoming frightened stampeded men and horses in his pitching and bawling, ending by running up against the shack box table and knocking the prop from under it spilling the whole meal so that most of the work had to be done over again.

The cook never knew how many men would eat at his wagon, especially at the noon meal. He might be required to feed ten, twenty or fifty men but he was always equal to the occasion.

Much has been written extolling the greatness and skill of some of the noted "Chefs" of our great hotels. They worked with every facility known to the art of cookery. But our old time wagon cooks, with their open fires of dried cow chips, sagebrush and cottonwood sticks could fry a steak fit for a king. Their pot roasts fresh from the Dutch oven were incomparable. Bill Anderson was known far and wide for the perfection of flavor of his bean soup. Gus Roberts sourdough biscuits were snowy and flaky.

The cook was the victim of many a practical joke, most of them taken in the spirit of fun, in which they were perpetrated. He lent a willing ear to every fellow's grouch or tale of woe. The horse wrangler, usually a young boy, was his helper when time from his other duties permitted.

Sometimes in camp at night, with a good fire going and a fresh beef on hand, we had a rib roast. The cook was not in on this, but every fellow cut out a rib or two full lengths. Stuck on a wagon rod or long pointed stick these ribs were held over the live coals and roasted. With a sprinkle of salt, they were eaten rare or well done, as liked, and the bones picked clean for they were delicious. Some of the Texas boys liked to rub a bit of the gall on their roast to give it flavor.

No nightmares followed these feasts for our digestion was perfect. Our days and nights in the open air gave us the most wonderful health. Almost never was a man sick. There were accidents, of course, especially during the summer months. I think it was the summer of 1886, that two men on the round-up were bitten by the little "hydrophobic" skunks. They went somewhere to have a "mad stone" applied to the wounds for at that time the serum treatment was not available.

It seems strange that no one was snake bitten, for we sometimes had them as bedfellows. One night we were camped at Wagon Bed Springs and had made our beds down on the bank not far from the springs. It was a moonlit night, almost bright as day. Sometime in the night I awoke and was annoyed by what I thought was a continual wriggling movement of the man who was sleeping with me.

In exasperation I asked, "What are you doing?" "Me," he answered, "Why, I'm not doing anything, I have not moved." "Well then," I replied, "there be certainly a snake in this bed, let us throw the covers back and make a quick jump." And, so we did. Not only was there a snake in our bed, but in the bright moonlight we saw dozens of them all about. Water snakes they were, out for a frolic in the moonlight, it seemed. At our movement they went scurrying back into the water and we lost no time in moving our beds back up on higher ground where we slept undisturbed the reminder of the night.

Considering the numbers of men engaged in the business of riding the ranges and the country traversed at all hours of day and night, often at reckless speed, the accident list was small indeed.

Part 2 • Chapter 16

Perry LaForce, a young Texan, sustained a broken leg by the falling of his horse while on the round-up. He was taken to the nearest place of the accident which happened to be the road ranch at the Crooked Creek crossing, kept by the M.C. Martz family. Here he was nursed by kindly Grandma Martz and her adopted daughter, Miss Belle, she later became the wife of Mr. Dave Mackey.

The Martz family were real pioneers coming west from Missouri in 1877. They traveled by wagon, driving with them a herd of about 100 head of cattle. They went into New Mexico and Colorado, but were forced to turn back by the lack of grass for their cattle as there was a drought in that part of the west at that time and so after many hardships they had settled on the Cimarron River. In 1879, they had moved to a place on Crooked Creek, built a large sod house suitable for a road ranch and many travelers were entertained and found food and shelter there. The Post Office of Odee was kept at this place for a number of years.

During the summer round-up of 1886, I sustained a painful injury that laid me up for a couple of months. We were out on the circle on the XI Range when my horse stepped in a hole and turned completely over. In the spill my right leg was wrenched out of place at the knee socket and pulled completely around so that my foot turned backward in place of forward.

Three of the boys, George Perry, Fred Powers and Smith Ruble, riding in sight of me came at my call of distress. They held a consultation over my prostrate form. None of them knew just what to do, but George Perry said that he had heard that to pull down and out on the limb in a case of that kind and the joint would spring back in place.

"All right then get busy," I said, "For this leg is no good to me this way." So they cut my boot away and then with Powers sitting astride my chest to hold me down, George pulled with might and main and Smith assisted turning the leg at the proper time and it popped back in place.

The sun was very hot so to protect me from it they placed three saddles, on end, and spread a saddle blanket across for a shade. One of them then rode to the XI Ranch for the team and buckboard and in due time I was carted into the ranch where I was nursed for several weeks by Pete Craig, one of the men working there. Later, I had a doctor come out just to make sure that I would not be permanently crippled, but the boys had done their work well and no further work was necessary.

As soon as I was able to be moved, I was taken to the home ranch. All the boys were out working, it was tiresome business just lying around doing nothing. Mr. Taintor finding time heavy on his hands, too, decided to go to Dodge City for awhile and took me along. We had been there for some time when one day a great excitement arose. All day long, settlers came pouring in from the south, on foot, on horseback, families in wagons, and all panic stricken. There was an Indian uprising, they said, the Indians were on the warpath. All sorts of wild tales went the rounds.

Toward evening Mr. Taintor came up to my room. "I am going out to the ranch," he announced. "I can't believe there is anything to this Indian scare, but if there should be I ought to be out there to protect my property. And just in case there should be, and anything happened to me, I am going to leave with you the address of my people in Connecticut so that you can notify them."

"All right, Fred," I answered, "only I am going out with you." And that was how I came to take the first ride after my accident. We had come up to Dodge City in the buckboard, of course, but Mr. Taintor had bought a couple of saddle horses during our stay and we decided to ride them out home so saddles were loaned us. I borrowed a rifle and he bought a new one so we left next morning at 10:00 A.M. with plenty of ammunition and rode in to the ranch. Having seen no signs of any hostile redskins on our way, the whole Indian scare was without foundation.

I must confess that my seventy-five mile ride after two months of enforced idleness was a bit wearying, but I was glad to be back "home on the range" once more and knew I could now get back to work and was happy to do so.

Chapter XVII

No Man's Land

At this time, the summer of 1886, the best and most carefree of the cattle ranching days were over, passed away forever. There was a very decided change in local conditions going on at this time. The blizzard and continued cold and snow of the previous winter had taken a great toll in livestock.

The loco weed continued to harass and worry us. That fall it grew in such abundance in our horse pasture that all of our horses were becoming affected by it and so we all turned out to what seemed the impossible task of cutting it out. We dared not leave it on the ground, but gathered it up and hauled it away to be destroyed.

Our outfit, the GG, and the XI leased a fenced pasture on Snake Creek in Clark County and placed there a beef herd of 600 big three and four year old steers to fatten for market. Instead, they all became addicted to the loco weed, as we found to our grief when we came to drive them to Dodge City. For we found them hard to handle, slow and stubborn, almost impossible to cut out, so that instead of shipping the brands separately as had been intended we loaded them in haphazard to be separated at Kansas City and we had a hard time getting them into the cars at all.

All the ranges were being cramped at this time by the homesteaders. Little towns were springing up here there and everywhere. Some of these towns only lasted a few weeks or months but a few became permanent.

The town of Englewood, in Clark County, was growing and the Santa Fe Railroad was being built into it that fall of 1886.

On the Cimarron River a few miles north of our ranch there was quite a settlement of "Dunkards." Fine people and good peaceable neighbors. A post office was established and a schoolhouse built, this was also used for church gatherings.

Some of the leading families in the Cimarron River community at that time were the Moore, Ennis, Whitaker, Roberts, Berends, Maphet, Brown, Fox, Meyers, Pietz, and Spurgeon's. On Crooked Creek, north of us, were the Parks, Pemberton, Shaefer and other families. Most of these families remained as permanent settlers.

There was never any real trouble between the settlers and the cowmen in this part of the country, although both had their troubles. Many of the settlers were unable to fence their fields and although we rode constantly to keep the cattle from trespassing, there were times when they do so and then there were damage bills to pay. Mr. Taintor furnished fencing for many of their fields and gave employment to many of the newcomers.

On the high flat, south and east of the ranch, the land was all squatted on and settled on for a season or two. Many of these were people of refinement, good people who tried hard to make homes there. But when the drought and hot winds of the following seasons came on, these people on the high lands could not stay. In the course of a couple of years they had all moved out. Only the ones along the steams remained and made permanent homes and not all of those. The one's who did manage to stick it out saw many hardships.

Prominent among the settlers on the Beaver River now were the Maple's, Beebees, Braidwoods, Evans, Grove's, Foshers and a number of others all of whom made permanent homes.

On the Kiowa were the Tafts, Judys, and Pettys, all of whom are still residents of the county.

None of the land in the "neutral" strip was open for settlement, some of it not sectionized as yet. The north tier was only lain off in townships, and so all the settlers were then "squatters."

When it became generally known that here was a strip of country 33 1/3 miles wide and 165 miles in length, over which was no jurisdiction of the law, it became a veritable place of refuge for criminals and lawbreakers of every description. Here one might kill, steal or commit any sort of crime and yet no law could touch him.

The City Marshall at Englewood, Kansas, once told me that he was holding one hundred warrants for the arrest of men supposed to be across the line in the strip. But in order to serve these warrants, the men wanted would have to be apprehended in the state. These men, however, were careful to stay where they knew they were safe.

It was a strange situation and some strange doings developed from it. A large "still house" was built on Hog Creek about ten miles southeast of our ranch. Here liquor was distilled in quantities and sold openly. People from different states came here to purchase the output.

It was a thriving business that went unmolested for some years. Near the still house was Neutral City. Here an attempt was made to enact and enforce local laws. In the end, it was mob law.

It was a time of much unrest, there were several shootings and hangings. The really law abiding element was not in favor of this sort of law enforcement.

There was talk of taxation, but it never went any farther than that. A town was started on the Beaver River and named Benton. It was to be the county seat town and here a newspaper, "The Benton County Banner" was published for a time by E. L. Gay, in the late 80's.

People of various vocations came to the strip, squatted on a quarter of land and tried to make a living, there were broom makers and boot makers.

On the Cimarron River, at the crossing of the Healy Trail, were an old Dutchman and his wife, both of them odd, shrewd characters. The old man's vocation was the making of rustic willow furniture, he did quite a thriving business in this for a time, and as a sideline he also made counterfeit money. Coming here with nothing but his team and wagon he prospered and soon had a nice little bunch of cattle.

The farmers, during the seasons of sufficient rainfall, raised some good crops of corn, cane and rice corn, but there was no market for it and it sold as low as 10 cents per bushel. And, then when the dry season and hot winds came on, the crops all burned up.

When the lower Oklahoma lands were thrown open for settlement there was a great exodus from the strip. We then lost some of our good citizens as well as practically all of the undesirables, for the latter were of the class that is never long satisfied anywhere and like best to be on the move.

A cowboy funeral, killed by his horse on a round-up.

Chapter XVIII

The End of Open Range

The winter of 1886-1887 we had another bad blizzard. On January 1, 1887, I took the team and buckboard and drove to Englewood intending to go to the C.O.D. Ranch, northwest of town, to stay over night. As on the previous year before, our first bad blizzard, this was an unreasonably warm day and I grew suspicious of the weather and on reaching the C.O.D. Ranch told the boys that I would not unharness my team but would go back home. And although they urged me to stay and laughed at my fears of bad weather I persisted and after a couple of hours visit started back home, going all the way back coatless.

On reaching home all were surprised to see me back for I had told them I would stay over night. My fears of the weather were not unfounded for at daylight next morning the storm was on and raged in unabated fury the entire day and part of the night. A terrible storm that again took its toll of livestock and some human lives as well, for many of the new settlers were totally unprepared for such extremely cold weather and there was much suffering, especially where people were living in little board shacks.

The loss in cattle was not so great as of the previous year, for the range cattle had been so thinned out that the grass was much better that fall, so cattle were in better condition, and too, this storm was not followed by so much snow and severe weather as of the previous year. But even so, it was a hard winter for the cowmen and we boys rode every day caring for all the cattle on the range as best we could but the loss was heavy.

The outlook for the cattleman was far from encouraging for the settlers were encroaching more and more on the range. Many of the larger ranchers had already moved out and some of the smaller owners had been put out of business by the loss of practically all they possessed.

Mr. Taintor had spent that winter in Connecticut and when he returned that spring we talked things over. It was decided that he would go to find a new range and I was to gather up the cattle and hold them in readiness to be moved elsewhere.

He went to the sand hill country of western Nebraska and to Wyoming in search of a location while I, with the outfit, worked every stream in the country. The Canadian River, Wolf Creek, Beaver River, Cimarron River and their tributaries in an endeavor to gather up everything we had left. The winter of 1885-1886, we had started in with 5000 head of cattle and now in the spring of 1887, we gathered about 1800 head, brought them in home, put them under close herd and branded the calves all in readiness to move out.

In the meantime the settlers on the Cimarron River, hearing of our intentions, came to the ranch several times to ask us about it, for they did not want us to leave. I told them they would have to see Mr. Taintor about that, so on his return they called a meeting at the schoolhouse on the river and asked us to come to it to talk the situation over. Mr. Taintor and I did so. He had been kind and helpful to these people and they now asked him to stay, promising to help him in every way possible. And that if he would furnish fencing for their fields, the cattle might run at large among them unmolested, and they kept their word for in the end it was decided that we would stay right here and if possible recoup the loss.

The arrangement, as formulated at the meeting, proved entirely satisfactory. The fields were fenced and line riders held the cattle within established bounds. The old careless, carefree days of cattle ranging had passed away. As gracefully as possible we adjusted ourselves to the new and changing conditions of the times. Looking back over the years I can now see that we did not, at the time, realize how much of a change was going on. For we were still just a bunch of young chaps full of tricks, always looking for fun and we were not now living the isolated lives of those first years.

At many of the ranches there were now women, the wives of owners, or of some of the men. For with so many young women coming in at the settlements, a number of the boys were now forsaking bachelorhood.

At the **M** Ranch on the Beaver River, Mr. Over had brought his bride. She was a young Frenchwoman, who very soon became a great favorite with all the cowboys because of her friendliness and jolly good nature. We called her "The Madame" or "Madame Over" and a visit to the Over Ranch always meant a hilarious time, for both Mr. And Mrs. Over were young, lively and full of fun.

On one of Mr. Over's trips to Dodge City he came across three young boys looking for work on a ranch and brought them to the **M**.

Part 2 • Chapter 18

They were known by the names of "Irish," "Beans," and "Queer." All three were city urchins, left to shift for themselves when mere children, and they would fight at the drop of a hat. Of course, this delighted the older men and they managed at all times to keep the kids scrapping. "Beans" and "Queer" soon drifted farther on, but Irish remained in the country to the end of his days. Born in Boston and brought up on the waterfront, as a very young boy he made several trips from Boston to Liverpool with shipments of live cattle for Armour and Company. Later coming to Chicago with Armour's, where he met many western cattlemen and heard so much talk of ranch life that he was eager to try it and so at fifteen years of age started to work on the **M** Ranch.

The life suited him well and here he made his home. Brimful of Irish wit in his more mature years the sayings and doings of "Irish" McGoven were quoted wherever a group of cowboys were assembled. He lived for a practical joke and one of these very nearly got me into trouble.

Meeting him on the street of Beaver one day he said, "Where are you going?" And I told him to the barbershop for a shave. With a wide grin, he said, "All right, go ahead and stick around awhile and see what happens." I asked him what it was he expected to happen, but he refused to explain, so I went on my way. In the shop a customer was in the chair for a haircut, so I sat down and talked. When the hair cut was finished the barber reached for his water bottle, gave the customer's head a generous sprinkle rubbing it in quickly with both hands. The effect was startling, to say the least, for Irish had had excess to that bottle, emptied out the water and replaced it with "hokey pokey."

The victim acted much in the same manner that a hokey pokeyed animal would. Leaping from the chair, yelling and pitching about, he lunged for the door, fell on hands and knees and then pitched headlong through the screen taking door and all with him into the street.

The astonished barber getting a whiff of the offensive odor of his water bottle now realized what had happened. Angrily he turned and accused me, flourishing a razor meanwhile. I had a hard time to convince him of my entire innocence of the whole affair. When I saw Irish again and related the occurrence he was highly elated.

The greatest social event of the spring of 1887 was the wedding at the H Ranch of Miss Elizabeth Schmoker and Thomas Johnston. The cowboys from far and near gathered there for the event.

"Slick" Johnston was "one of the boys," born and reared in Cleveland, Ohio. There the Johnston family was neighbors and close friends of the Rockfeller family. He came west with Frank Rockfeller when the Rockfeller family established a ranch first in the Indian

Territory and later the ranch in Barber County, Kansas. After several years on the Rockfeller ranches he came on to this country with the A H outfit owned by Fred Hobbs.

They located on Crooked Creek and ranged there until the settlers came in such numbers that a move was imperative, so the cattle were taken to New Mexico the fall of 1886. "Slick" as foreman of the outfit helped move the cattle to their new range and then returned to this country and was with the **M** outfit at the time of his marriage. The newly wedded pair came to the ranch to live for a time and was accorded a royal welcome.

They later established a home of their own on Horse Flat, a couple of miles west of Taintor's Ranch, afterward moving to their homestead on the Cimarron River. The Johnston home was "home" to all the boys of the range country.

When "Slick" brought his bride to the ranch to live for a time, our old cook, Major B, was retained as round-up cook. The Major certainly deserves mention in this history of ranch life in the early days. Born of a fine old Virginian family, he had unfortunately became too much addicted to drink and in order to avoid bringing disgrace on the proud family name, he came west and drifted into service in the regular army where he served for some time. He had been with the command that followed the Northern Cheyenne Indians in 1878, when they made the break to return to their former home in the North. He was present at their final surrender, when trapped and cornered in a deep canyon, many were killed and the remnant of the once fierce and warlike band were overcome by the superior strength and resources of the white soldiers.

His army service over, Major, whose title was merely by courtesy of the cowboys, now came to the laxer life of the cow camps and added there to a personality that was outstanding. A great reader and an ardent Democrat, we had many a heated argument. Older than most of us at the ranch, we boys teased him and played tricks on him and it was perhaps in retaliation that he developed the uncomfortable habit of going out on the hilltop on afternoons when he had idle time to pass away, there practice rifle shooting.

He was a good marksman and many a time on our way in from line riding, the Major playfully dropped "pot shots" all around us and the rifle bullets whistled by all too close for our comfort or safety. And, not only did he pull this stunt on the ranch boys but strangers riding in were also sometimes accorded this doubtful welcome. At other times, instead of rifle practice, he went to his rendezvous unarmed and went through a series of army maneuvers, marching and counter-marching,

Part 2 • Chapter 18

rasping out his own orders and beating time with a stick on a small tub, in lieu of a drum, and this gave me my chance to give the Major his innings.

He was busy with his maneuvers one afternoon, when I got home and seizing the rifle I went outside. Taking careful aim, my bullets kicked up the dust all around old Major. He yelled and gesticulated, but I kept right on until I saw him fall. When for a horrid moment or two, I thought perhaps I had hit him, but as soon as the firing ceased he rolled over the hill and came down the little canyon to the ranch on the double quick.

"See heah boy," he announced, "this damn foolishness has got to quit." "Sure," I agreed, "that's just why I have been giving you a dose of your own medicine, you have been pulling this sort of thing on the rest of us for a long time. Now see that it does quit!" That ended the gunplay. But the boys continued their tricks. Once when he was outside for practice one of them fastened a pail of water over the kitchen door so that when he entered it he was deluged.

We had all noticed his habit, when dishwashing was finished, he would set the pan of dishwater on the extreme edge of the stove, steadying it there with one hand while with the other he reached out and opened the door preparatory to emptying out the greasy mess. So one night it was hastily arranged to have a bit of fun. The kitchen was Major's domain and it was understood that he did not like to have anyone underfoot when he was at work there, but this evening Jimmy Givens strolled in while the dish washing was in progress and started a conversation that soon led to a hot political argument during which the Major became quite oblivious to any outside noises. It was then that the boys placed a long, heavy post against the door, in such a manner that it would fall inside when the door was opened. Still absorbed in his argument Major set his pan of dishwater on the extreme edge of the stove, steadying it with one hand he pulled open the door with the other when, "Bam" the heavy post fell in, knocked him down and the overturned pan of dishwater landed squarely on top of him.

I was on the outside watching through the window and saw that the Major had sensed instantly the whole scheme, for he was up like a cat and at Jimmy the conspirator, pummeling him with might and main. By the time we ran around to the door and got in there to pull him off, he had Jimmy pretty well skinned up. Major was quite wrathy for awhile, but soon cooled off for at heart he was just a boy and enjoyed fun, when it was at the other fellow's expense.

When the settlers came in on the river near us, the young boys of the community found the ranch an attractive spot. Often we all were gone for weeks at a time on the round-up, leaving Major in charge at the ranch. It was at such times that he delighted to entertain these youngsters, feed them, clothe them in the best of the clothing we had left behind, and send them happy on their way.

Returning home we searched in vain for extra suits of underwear, boots, etc. At such times I would ask, "Major, where are my boots and that coat and vest I left here?" "See heah, boy, how do you expect me to keep track of all your things, I don't know anything about them." And he went on with a great show of innocence that was the sure sign of his guilt. Invariably some time later, we chanced to meet three urchins wearing boots several sizes too large for them and with coat sleeves turned back several inches to permit use of their hands.

This misfit clothing had a strangely familiar look and the boys when being questioned said Major had given them the things. Back at the ranch Major being confronted with this evidence always came back at us with, "Now see heah, boys, you are plenty able to buy you some more clothes, you don't want to see these poor little fellows going without, of course, I gave them those things."

But when the longing for drink came upon him as it did at intervals, he always managed to get it somehow. Once when we all left for a short time and he had neither team nor saddle horse at hand, he took his coat over his arm and carrying a small coil of rope, walked down to the settlement on the river and going to the home of the "dunkard" minister of the community. He told a tale of having heard of a stray calf belonging to the G G that had been down on Horse Creek and as everyone was gone from the ranch he thought he had better go get it and would Mr. E (the preacher) lend him his team and wagon and let the two young boys go along to help him? Mr. E was a good friend of ours and he readily assented, glad of the chance to do us a good turn. The boys were youngsters about eleven and twelve years old and they and the Major drove happily away.

The following evening on my way home I chanced to come by Mr. E's home and was met by a much worried father who told me of Major's errand there the previous day. And how, as yet, neither he nor the boys had put in their appearance with the troublesome "calf," I knew at once how to interpret the Major's story and I said, "Mr. E, Major did not go to get a calf, he wanted to get to Horse Creek where he knew he could get Moonshine whiskey and he invented that story as an excuse to be taken down there. But don't worry, he will see that no harm comes to the boys. When he gets his fill of whiskey he will bring them back all safe."

That evening they drove in to the ranch, the boys driving the team, Major sitting in the back of the wagon, weaving and talking in a hilarious state of inebriation. The preacher's young sons, a thoroughly frightened pair of small boys, unloaded their passenger and made for home. We at the ranch felt rather disgraced by this stunt of the Majors. We felt that he should have at least left the youngsters out of it.

The T6 outfit of R.K. Perry.

The last T6 trail. The herd was gathered from the Beaver and the Cimarron and is being started for the railroad at Meade, Kansas.

Crooked L outfit, ready for the round-up circle drive.

Chapter XIX

The Gray Wolf

By 1887, many of the large cow outfits had moved out. Among these were the Beverly Brothers, S. S. Johns, The Day Outfit, Hardestys and a number of others. Among those that remained where the XI's, the Crooked L's, (T.6) owned by R.K. Perry, and K.K. (Healy Brothers.)

The roundups were not so large now and so we did not have to work so far out from home. My outstanding recollection of that year's round-up was that of July 4th, one of the hottest days I ever saw.

We were working on the Kramer Ranch then, four miles west of Beaver City. Even at daylight it was unusually warm and the sun came up with such a fierce heat that we had to proceed slowly with the work as we made the circle. Throwing the cattle all in on the river, we did not, as was customary, hold the round-up before going in to dinner, but left the cattle in the water while we went to the wagons for rest and food. We were hoping a breeze would spring up in the meantime, but in this we were disappointed. However, we had always gone on with the work regardless of weather so soon after our meal we held the round-up, working slowly because of the intense heat.

About four o'clock cattle began to die. An animal would walk off a few paces, quiver for a minute and fall over dead. We turned the cattle loose and let them drift back to the river, some died on the way, others after they reached the water. In all, about fifty head of cattle perished from the heat that afternoon and evening.

We were working in a sandy country and the bald sand dunes to the north of the Beaver River seemed to catch the heat and glare and reflect it over on us where we were at work. The heat held until long after dark that night. None of the men suffered heat prostrations, though all felt the effects of it, so that for a day or two after, the work lagged.

That summer was rather a dry one. The crops of the settlers were short. Most of them had a little early corn and the women dried corn for future use. The wild plum crop was an abundant one. Having little in the way of canning facilities and sugar being scarce, plum leather now came into favor. This was made by rubbing the plums, cooked or uncooked, through a colander. The resulting jam unsweetened, was then spread on plates and platters and placed in the hot sun and wind where it soon dried on the top, then it was turned over and allowed to dry on the other side. In three or for days it was thoroughly dry and then the red-brown discs were stored away in flour sacks for winter use. When wanted for use later on, one or more of these discs were soaked over night in water enough to cover, cooked and sweetened. The plum leather jam was very good.

Jackrabbits provided meat, in many cases, the only meat available. Rice corn, a drought resisting grain, was then introduced. The grain was smaller than kaffir corn that later supplanted it, and it was much harder. The stalk, of the rice corn, grew four to five feet high and it too was hard and woody, so the fodder was not a good one for their milk cows, but it had to suffice. In their extremity this rice corn grain was used for food. Ground fine in a feed grinder the meal was used for bread and pancakes. The bread was dark, hard and solid, but the pancakes were not so bad. Many of the families ground the daily supply on the coffee mill. When coarsely ground and cooked it served as breakfast food, parched it was used as a coffee substitute, and popped its tiny grains made a delectable popcorn.

Most of the families had a milk cow or two and a few hens. So they managed to exist and though the fare was monotonous and very, very plain these people, for the most part, enjoyed the best of health.

If the day was a hot one, the herd lingered on the water until nearly sundown, while the patient guards stood in somnolent attitudes scarcely moving for hours on end. If nothing appeared to mar the quiet scene all went well. The herd returned, each cow finding her own calf now ready for its supper after which all spread out for quiet grazing while the thirsty guards made their way to the stream for the long delayed drink. But if, as sometimes happened, an enemy appeared upon the scene, the cows in charge were instantly on the job of protecting the young.

Sometimes when riding the range our attention would be attracted by the frantic bawling of cows. Rising to the scent of the trouble, we usually found a number of cows in a circle. Within the circle were their calves and a short distance away a pair of gray wolves circling the bunch looking for an opening. When a wolf ventured too near, an enraged cow sprang to meet him. Sometimes at night the bawling of the cows became so insistent that we knew there was trouble abroad and then we rode out and with rifle shots and shouts managed to rout

Part 2 • Chapter 19

the killer for the time being, but they always got their fill in the end anyhow.

The ranchers made determined efforts to wage war on the pests. Most of us carried rifles on our saddles and regular trappers were employed at many of the ranches, but the gray wolf is very wary and few were trapped or shot.

It was about the winter of 1887, when the wolves were at their worst, that Mr. Taintor brought to the ranch Pete Holtz who claimed to be an experienced trapper. So we fitted him with horse, saddle, bridle, rifle and strychnine for his work of exterminating the grays.

Pete was a real character, short, red-haired with a testy temper that flared up unexpectedly. He had a fund of odd humorous remarks and a picturesque profanity that proved highly amusing to the bunkhouse bunch and there were few dull evenings when Pete was about.

He worked hard and faithfully putting out a string of steel traps as well as strychnine baits. For these he took planks, sawed them in short lengths in each of which he bored several auger holes about an inch in depth, into these he poured melted tallow containing strychnine. He claimed the wolves would lick the tallow and so get the poison. These planks were then nailed to stakes driven into the ground.

In all his preparations he wore gloves so as to leave no human scent on the bait. In preparing the tallow and strychnine that had to be mixed over the fire on the cook stove, he was often so careless that we feared for our lives.

For many weeks Pete rode the range and worked with traps and bait. Often coming in at night with cheerful reports of success in his work, but never a gray wolf did he bring in to back up these reports.

However, we never lacked for venison while Pete was with us. One evening we came in and found him busily engaged in skinning seven deer that were suspended from tree limbs. He had shot them up the creek, came in and got team and wagon and hauled them in. As a hunter of wild game Pete qualified all right, I could not say as much for him as a trapper of gray wolves.

Every night after supper a card game was started in the bunk room and it was then that Pete's temper was tried to the utmost. The boys knowing his weakness delighted in getting a display of it by deliberate cheating and standing in together against him so that night after night Pete quit the game mad as a hornet. But it was a much smaller matter than this that caused Pete to leave us in a huff.

One night after most of us were asleep, Joe King and Don Holman took Pete's trousers and mine to the kitchen, cut every button off of his trousers and sewed them on mine. Next morning the boys awake and watching when Pete got out of bed, pulled on his pants and fumbled for the buttons that had all been there the night before, but now, "Nary a button."

"Why," Pete snapped, "There's not a button on my trousers."

I had slept in an adjoining room and just then walked in demanding, "Who the devil sewed all these buttons on my pants?" For, I had a double row of them on mine.

That was too much for Pete, instantly he put all the blame on me, though I was as much a victim of the trick as he was, but he was too angry to see that. From a trousers pocket, he fished out three or four eight-penny nails and used them to fasten the trousers. He ate his breakfast in gloomy silence and that day brought in his traps and without a word to anyone walked off and left us.

It was a long time until I saw Pete again. When I did he was cooking for the T outfit and the button episode still ranked for he greeted me very coldly. However, in time all was forgotten and forgiven and we were once more the best of friends.

But, to return to the subject of the gray wolves, they left as suddenly as they had appeared. For they do not like too much human companionship, so when the rough hilly country became invaded by the rush of settlement, the wolves migrated to less settled regions.

Many of the boys, while out early in the morning line riding, reported seeing them on the move in groups, always traveling to the southwest. I counted twenty-five of them in a bunch one day going down a deep draw on the Healy ranges.

Coming upon them quite unexpectedly, I hastily drew my rifle and fired a number of times, but failed to get a single wolf. Another man farther up the river had much the same experience. Riding up to the top of a long ridge, he saw in the draw below it and just ahead of him, sixteen large gray wolves traveling steadily southwestward. Being alone and unarmed, he said he felt a cold chill creep up his spine at the sight. So without attracting their attention, he quietly dropped from sight on his own side of the ridge and let them go on unmolested.

And so the main body of the destructive creatures moved on, only a few remained in these parts. In time these were driven out or destroyed, for they were hunted relentlessly. A few were trapped, several were roped and dragged to death.

One female, for several years, denned and raised her young in a rough canyon on the north side of the Cimarron River, but with great cunning she invariably made all of her kills over on Crooked Creek, thereby throwing hunters off her track. But one spring we found her den, high up on the side of the rocky canyon wall, a hole into which a man could crawl.

One of the boys went in and brought out the six whelps it contained and they were killed. She then disappeared and was never again seen here about. And no more calves were killed on Crooked Creek, eight or ten miles to the northwest, where she had found her favorite hunting ground. So it is likely that she too moved on to some less settled region. And, it has been many years now that any gray wolf has been seen in these parts.

"Slick" Johnston and family at their house on the Cimarron.

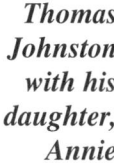
Thomas Johnston with his daughter, Annie

Fred Taintor GG outfit taken on the Cimarron River at the Tom Johnston ranch home. The little girls are Anna and Mildred Johnston and Laura Mackey. F.M. Steele photograph.

Chapter XX

The Prairie Fire

The old time hunters, the Indians and the cattlemen were very careful to guard against fires, so it was that during the first ten years of my life in this country there were no fires of any great consequence. However, I was told by some of the old plainsmen that some time in the seventy's there had been a fire that swept from the Smoky River in Kansas, to the South Canadian River in Texas. It also covered quite a wide strip of country.

Carried by a gale of wind from the north, this fire crossed all the streams between these two rivers. The wind was of such violence that burning tumbling weeds and other light stuff was easily carried across the streams and sand draws.

The worst fire in my experience occurred in March 1888. It was about that time that squatter settlement in No Man's Land had reached its zenith. Along the streams, over the high flats, in the rough hills anywhere and everywhere might be found dugouts, sod houses, or rough shanties of those early settlers.

Most of these people were having a hard time making a living for there was little work to be had and the crops had been poor. Those who did have a little grain for sale could find no market for it, except that which the ranches afforded and in those years only a few saddle horses were kept and grain fed during the winter.

Ten or fifteen cents a bushel was the price for cane and rice corn, for when hauled to town it could not be sold at any price for no one considered these grains fit for horse feed, but corn was used and it too was very low priced.

So while these farmers had plenty to feed their poultry flocks, the eggs they produced were worth but five cents a dozen and often could not be sold at all.

Naturally such conditions made much dissatisfaction and as the cattlemen were prospering much of this dissatisfaction and ill will was directed against them. There was considerable talk of burning off the ranges to make the cattlemen move out.

The great majority of the settlers, living on what we claimed as the range, were our friends and we had no trouble with them. But over on the Beaver River an old radical had moved in that fall and while we were on beef round-up over there Old Major, our wagon cook, drove close to this man's shack one evening and proceeded to make camp.

When I rode in to the wagon a short time later, he and Major were having a heated argument for he had ordered Major to get out of there, saying he did not want a d___ cow nor a chuck wagon on his place.

"We are organized," he shouted, just as I rode up. "Then I will dis-oahganize you right now," Major shouted back at him. And forth with, shoved the old black coffee pot he was holding in his hand into the man's face, blacking it in the most ludicrous manner.

Major then went on with his supper preparations, while I talked to the fellow and explained that we were merely camping there for the night, and had just as much right to be there as he had. For none of the land was open to settlement, no one, not even the U.S. Government seemed to know just whom this land belonged to at that time, so he calmed down. But he was a born troublemaker and that winter began advocating setting fire to the range.

Some of our friends among the settlers told us of this threat and we assured them that if this were carried out it would be the settlers, themselves, who would suffer most of the consequences for we could move the cattle out.

So on a windy March night with a gale blowing from the southwest, this man set out a fire, and what a fire it was. It started in the tall sand hill grass on the north side of the Beaver River. The demon flames gathered momentum as they swept up through the hills and out onto the high flat between and Beaver River and the Cimarron River. The settlers were helpless in its path.

At the Ranch we soon saw the reddening skyline. There were seven or eight men and we saddled our horses and rode out to see if anything could be done. To have faced the head fire in that wind would have been suicide, so we rode to the sidelines and even there we were trapped and had to dash through, such was the rapidity with which the fire spread.

It was an awe-inspiring sight and it was quite evident that we could do nothing to stop it, so we rode to some of the homes that the flames had not yet reached. Here we found women and children crying with fear and men stupefied at the sight, realizing how powerless they were in the path of that fire.

Part 2 • Chapter 20

The smoke was driven ahead in a dense choking mass, behind it the flames came with the speed of a racehorse. At the homes where we stopped we told the men to picket their horses and milk cows out on plowed ground. But at most places these were in sod barns with thatch roofs and these perished or were badly scorched.

Fortunate indeed was it that the homes were all of sod or dugout with dirt roofs, else all would have met a common fate. At one of these homes where we stopped there was a small stack of four or five tons of fodder. By the combined efforts of ten men we managed to save this stack of feed. It was the only one left on the flat after the fire had swept in, but we had a hard time saving it, for the wind picked up, burning cow chips and flinging them far ahead, these and the sparks that carried far constantly set out new fires.

When the canyons were reached, tongues of flames raced down the ridges. Here many places were left untouched, the fire ran down to the river's edge where it finally burned out.

As we had warned them, the settlers themselves, were the one's who suffered most from the fire. Most of them, of course, had not been in favor of it, but here and there had been some who thought it might not be a bad idea and now every vestige of grass and feed had been destroyed. Most of their livestock burned, or at least badly singed, the whole country a charred and blackened mass with here and there a soddy or a dugout as landmarks in the utter desolation.

We rode out next morning to find if any cattle had perished, but the wise old range cows had sought refuge in the sand draws and deep canyons where the fire leaped over.

The settlers were dismayed by their predicament. A meeting was held and it was decided that the man who had set out the fire should be hanged. A delegation was sent to the ranch with word of what was proposed and we were asked to participate. This we declined to do. We pointed out to them that it was their fight, not ours, since they were the real losers. We had escaped with only the loss of some fence posts out in the horse pasture. There was still grass for the cattle north of the river, and spring was not far off.

There was much talk and many threats against the one who had set out the fire, but they lacked a real leader so while the discussion was still going on the culprit prudently made a hasty move out of the country.

There was never anymore talk of setting out fires to drive out the cattle. Sometime later though, the cattle owners found a new cause for worry and it came about in this way.

The settlers in a determined effort to find some crop suitable to the dry climate thought they had found a solution of the problem when it was declared that the Castor Bean could be grown here with very little rainfall.

Accordingly quite an acreage of these ornamental looking plants were put out during the next couple of years, but the result was only another disappointment.

For while the plants grew and thrived, for some reason the crop was not a remunerative one and in the fall after the beans had formed in clusters at the top of the stalk, it was found that they were poisonous for cattle if eaten in quantities.

It was during the fall round-up that this fact was brought to our attention, for most of the fields were unfenced, some of them abandoned and in these the cattle roamed at will. We were much mystified at first, when cattle being brought into the round-up dropped out here and there and in a few minutes died. On examination it was found that in every case the stomach was filled with castor beans.

The settlers themselves lost quite a number of their cattle in this manner so the crop was cut out by all except a few with a grudge, fancied or real, against the cattle owners and these few now planted them merely for spite. But in the course of a year or two, these settlers holding a grudge, all moved out of the country and the castor bean, as a crop, became one of the things of the past.

The winter of 1888 was a particularly hard one for most of the squatter settlers. Many of them were without means to buy needed food and there was no work. In some parts of the country the men left the women and children alone in their dugout homes and went east to find work if they could, or get aid in some way. But for those living on the range, Mr. Taintor felt a responsibility.

They had shown a spirit of good will and neighborliness, had cooperated and made it possible to continue in his business where others had been forced out. So now although he was not at the ranch that winter we had orders that these people should not suffer for want of food.

So at his expense those families were supplied with floor, coffee, sugar, beans, potatoes and beef. We killed twenty-five beeves that winter and the meat was distributed.

Our method was to drive the animals to be slaughtered to the center of a community, where the meat was to be divided. We roped and threw the animal for the men who then killed and dressed it, taking back home his share of the meat.

Part 2 • Chapter 20

At one place a few miles from the ranch there was living at that time a family of forty-eight persons. It consisted of the father and mother and a large family of boys and girls, all of whom had married, so with their wives and husbands, the children and grand-children made up the number afore mentioned.

Many dogs were about, the only asset possessed by the clan. Being without clothing necessary for the winter weather, some of the men of the family had gone down to Fort Supply and made known their plight, and the soldiers had responded generously so all were clothed in donated raiment of the U. S. Army.

They were in need of food also, so I rode by there one evening and told them to be ready next morning and I would bring them a beef.

Accordingly next morning I drove a cow over for slaughter. On nearing the place I roped her and the men, eight big husky chaps, came out to meet me. I had a good roping horse and could easily have thrown the animal for them, but I thought they needed some exercise and I some fun. I called them to grasp the rope and throw her and they responded with a will. The cow, a mild one, ran and fought, but they over powered her by sheer force of strength and number. The show had to be seen to be appreciated, but I certainly had my money's worth of fun that morning.

It was one of the boys, belonging to this family, who went out one night and stole a quantity of wire fencing belonging to a widow living in that vicinity.

As I have before mentioned an attempt was being made at local law enforcement, and this miscreant was caught and brought before the Justice of the Peace of Neutral City. He was found guilty and fined $5.00 and costs, but he had no money to pay the fine. As there was no jail, what to do with the offender was the question.

Someone suggested that he be thrown into an old deep dry well. In the end the matter was disposed of by putting him on the block at the front door of the Justice of the Peace's office and there sold to the highest bidder.

He was bought, on credit, by the owner of the large distillery on Hog Creek, not far away, and taken to the "Still" house and put to work, but work was not in his line. He immediately proceeded to get good and gloriously drunk.

When court convened next morning the "Still" man brought back his purchase. "Your Honor," he said, "I can't use him, he don't do a thing except drink all the whiskey he can hold and is therefore absolutely useless in my business. I am returning him to the court."

Now this unforeseen circumstance was a bit staggering and while the situation was being discussed the cause of it was left unguarded and ambling about aimlessly, or apparently so. He moved quietly around the side of the building, darted in back of another building and then was off across the prairie running like a jackrabbit being pursed by a pack of hounds.

Like a rabbit, too, he dodged now this way, now that, when the officer sent a few pistol shots after him merely to speed him on his way. When last seen he was still running zigzag fashion across the far horizon in the direction of home. On reaching home, he was hid out for some time, but no further effort was made to apprehend him.

Some time later I met and talked with the father, a devout old shouting Methodist, and he was feeling much disgraced by the affair. Soon thereafter the entire clan left the country.

It is not my intention here to include a full report of the efforts made at that time to establish a local government by the people of No Man's Land. That is a matter of history. But this Strip of country was named Cimarron Territory. Quite appropriately named, I would say, since "Cimarron" is from the Spanish and it's meaning is given as "out-cast, outlaw, or wanderer." Most certainly it was an "out-cast" for many a year.

The people at that time had elected a full set of officers and an attempt was made to receive recognition from the Federal Government in Washington. But it was only after five years of determined effort on the part of the leaders that No Man's Land, the orphan child of the United States, was adopted and given a home, so to speak, amongst its contemporaries.

In the meantime, the local courts lacked any real authority or dignity, and many of the proceedings were a mere farce, a travesty of justice. It is generally conceded that the recognition finally accorded the "Strip" was hastened by the "Hay Meadow Massacre" on the Dudley Ranch on the Beaver River. Several citizens of Kansas, camping one night over the line in No Man's Land, were followed and wantonly murdered as they slept, because of enmity resulting from a county seat war in Stevens County, Kansas. An atrocious act, which could not be overlooked, and it was found that no court had any authority to try these men for the murders.

One of the first acts of authority by the Federal Government after authority was finally established was the destruction of the large distillery on Hog Creek.

Our ranch was always a stopping place for all transients, so it was nothing out of the ordinary when one evening four men drove in to stay

Part 2 • Chapter 20

overnight. One of them was Wiltz Brown from Meade, Kansas, well known to us, and the others were strangers.

During the evening's talk no mention was made of their mission in our part of the country, and on leaving next morning they merely inquired the way to the distillery.

Their first stop was made at Neutral City at the saloon. One of the strangers asked the proprietor if his saloon was licensed. "No," was the reply, "there is no one to pay a license to."

"Then you may pay it to me right now," announced the Federal man and displayed his badge. The license was paid without parley.

The men then drove on to the distillery where they destroyed all the machinery, and emptied out the mash, but allowed the owner to keep all the whiskey he had on hand. And so ended, what for a time, had been a very lucrative business in No Man's Land.

The squatter population was rather a shifting one during those years of waiting for the opening of the land for homestead filing. It was not until early in 1889, that the main exodus took place, for it was April 22, 1889, that the grand opening or rush for homesteads took place in lower Oklahoma. It was then we lost all of our transient population, as well as many of the people who had come here with the full intention of making permanent homes and had found conditions too hard to be able to hold on any longer.

All of the high flat between our ranch and the Beaver River was now vacated, not a single settler remained. Most of the little towns too disappeared almost magically. Only along the streams where the settlers had small herds of cattle to depend on were any left and even there the numbers were considerably thinned out. For the ranchmen still remaining, there now was added opportunity for development and prosperity.

The Autumn of 1889, two ranch owners made arrangements with Healy Brothers so that they might move all their cattle on the Beaver River for the winter as the grazing was fine and abundant there. The Cimarron country was grazed short and some trouble had developed from cattle drifting up into Kansas as not many of the Kansas homesteaders had gone with the lower Oklahoma rush.

Line camps were established on the Beaver River for there were plenty of vacant sod houses now. Line riders held the cattle as closely as possible to the home range during the winter following.

From this time on, the ranch maintained a branch camp on the Beaver River with headquarters on the Cimarron River at the home ranch on Taintor Creek.

In the summer of 1890, the cattlemen of the Cherokee Strip were ordered to vacate that country before a certain date in the fall. This gave them very little time to move, which caused some of them to sell their herds at a sacrifice. And so it was that Mr. Taintor bought the K.H. herd of 1400 head of cows and yearlings, with calves thrown in, at eight dollars per head. Horses, mules, chuck wagon and all were bought at a bargain for this was one of the outfits forced to move from the Cherokee Strip and had no where to go. The calf crop when sold paid for the cows.

The RS Outfit moved out, but I do not know where. Another outfit bought out the Laurel Leaf Ranch on the Canadian River and moved their herd of twelve thousand there.

On leaving the home ranch in the Strip this outfit set fire to the rambling old ranch house and the barn for much bitterness was felt over what was considered an injustice. This evacuation came at a time when prices were at the lowest ebb and the time given for the move was so short it worked a hardship on all of these ranchmen.

More time might well have been given them, for after all it was not until three years later that this land was opened for settlement and the Indians were thus deprived of the rental the cattlemen had been paying for the use of the land.

In this interval when it lay idle it was surreptitiously grazed and the Indians received no benefit. Some effort was made to keep out the so-called 'boomers' but their efforts were, for the most part, not very effective.

The owners had many miles of good wire fencing and this too had to be sold at a sacrifice. Mr. Taintor bought twelve miles of this fence. A force of twenty-five men were employed to take up this fence and when brought home it was immediately rebuilt north of the Cimarron River as a line fence to keep the cattle from drifting up into Kansas, thus doing away with all our line riding to the north of the ranch. We now had a better and wider range than ever before and began to expand accordingly.

It was a time of prosperity in the cattle business. For while only a few of the large ranchmen still remained in No Man's Land, the day of the smaller ranch owner had arrived.

The great rush of home-seekers was over. It had been proven that farming alone was out of the question. The settlers now remaining, without exception, turned to stock raising. More attention was given to winter feeding of hay and forage crops.

Quite a number of the cowboys had married and were now living on small ranches of their own. Better homes were being built and many fields and small pastures were fenced. These changes came so gradually that one is scarcely aware that changes are going on.

N.J. Rhodes' ranch on the Cimarron.

The Hightower family at the OX ranch on the Beaver in "No Man's Land."

Chapter XXI

A New Experience

It was still necessary to hold the spring round-ups, although we did not now cover such a great territory as was included in those first ones. So the spring of 1891, we started on the South Canadian River and came across to the head of Wolf Creek. It was there that we had an experience never before included in any of our round-ups, and this was a twister, then called a cyclone.

We were all in camp waiting for the southern men to get in. As usual this time was spent in horse racing and various games and on this afternoon we were all lounging about camp afoot. The horse wrangler was herding the horses down on the creek bottom when we saw the storm coming from the southwest.

It was a most unusual looking cloud, none of us had ever seen one just like it and several ventured the opinion that it might be a cyclone, while others said, "Shucks, it's nothing but an ordinary storm." But when we saw the whirling, twisting mass that was gathering up everything in it's path, consternation seized the erstwhile careless cowpunchers.

Forgotten were the card games as well as the tin plate and the top that had been spinning in it at a nickel a spin, obviously the time for action was short. We yelled and motioned for the wrangler to bring up the horses, but many of the boys did not wait for them and now developed a surprising flair for foot racing. The horses came in with a rush and those of us remaining in camp caught our mounts in record time and started to get out of the way of the twister that was coming straight down the creek.

Directly across the creek, not far from our camp, was a small frame house, the home of Rock Fuller, who was on the Canadian round-up. Mrs. Fuller was alone in their home and just as we were leaving our camp I chanced to see Mrs. Fuller out in the yard walking about, crying and wringing her hands, seemingly too frightened to know what to do.

I turned and rode back to help her if I could, but a long marshy bog intervened. There was now a dead calm and I shouted to her, "Get into your dugout quick, shut the door and stay there." She heard and I saw her enter the dugout. Then applying the spurs and none too soon, up over the rim rock we raced in a mad scramble for safety, the worst frightened bunch of "punchers" that ever rode the range. Huge hailstones pounded us, the horses frantic under their blows and the rowels that hit their sides causing them to leap up over the rim rock like mountain goats. So hazardous were some of the ascents, made under the stimulus of fright and excitement, that we did not attempt coming down over some of the seemingly unscalable rocks that the horses had climbed in coming up.

Up out of danger's way we halted and watched the twister as it wrecked it's fury. We saw the Fuller house torn up and carried away and great cottonwood trees twisted out and taken along. Everything in its path seemingly was being snatched up into the black whirling mass. Not quite everything though, for when we returned to the wrecked home of the Fuller's we found Mrs. Fuller safe and unharmed in her dugout. The house and its contents had vanished, all except a small stand table that stood just as it had in the house. The floor it had stood on was gone, but the cloth cover on the table was undisturbed as well as a lamp, a few trinkets and a ring, such are the vagaries of these storms.

Our wagon, by some miracle, was left untouched. And we spent the next day following the path of the storm for some distance, gathering up clothing, bedding, furniture and parts of the Fuller house.

In later times the queer freaks of the tornado became well known. But at that time these phenomena were quite unknown to us, so it was something to marvel at when we found straws and feathers driven deep into trees or fence posts.

Across the flat were great holes bored into the earth where the trail of the twister had dipped down and these holes were as smooth and symmetrical as if carefully made by the hands of man instead of being the result of a frenzy of nature.

In our search for the debris of the Fuller house we spread out fan-wise and afterward around the campfire each had a story to tell of the work of the storm. The stories lost nothing in the telling.

"Sourdough" Charley, who had been a wagon cook for some of the old outfits at an earlier day, had now settled on the creek on a place of this own. He had come into our camp and listened quietly while the different tall tales were being recounted and then he told his story.

"At my place," he said, "there was a large iron kettle setting out in the yard. When the storm had passed on and I went out to pick up what

Part 2 • Chapter 21

I could find, there was my kettle, some distance away. It looked sort of queer and when I examined it, darned if that kettle wasn't turned wrong side out, clean as a whistle. The three legs that it set on were now on the inside and the sooth inside of the kettle was on the outside, blamedest thing I ever saw." Someone handed him a sack of Bull Durham and a package of cigarette papers and said, "Here Sourdough, you win." For his was the tallest story of them all.

Several others with their round-up wagons were camped up the creek from us and these outfits had experiences somewhat similar to ours, but there was one incident we all thought remarkable. A short distance from one of the wagons, three trees stood in a row. One of the men, Jim Beard, tied his horse securely to the middle tree and then lay down under a wagon as shelter for a nap. He was sleeping soundly when the other men saw the storm bearing down on the camp.

Just as everyone started to run from it, someone thought of Jim, rushed back and shook him yelling at the same time, "Get out of here there's a cyclone coming!" Jim dazed with sleep never once thought of his horse, but raced off on foot out of harm's way. Returning after the storm had passed, he found his horse unharmed, still tied to the tree, but the trees on either side of it had been uprooted and carried away.

Several weeks later, working at the mouth of Camp Creek, on the Kiowa, we experienced another twister. This time when we sighted it coming over our way we knew what to expect and prepared for it, that is, those of us who had been in the previous one did. We had the wrangler bring up the horses and eight or ten of us caught and saddled our horses, while the others not having seen the previous storm refused to become alarmed, and only hooted at us. But, as it approached nearer and the awful appearance of it became real to all, they yelled loudly for horses, "bring up the horses." Those were quickly brought up once more and now every fellow ran in with looped rope to catch anything available for there was little time to spare.

It so happened there was much tall sage brush at this particular spot and now the dragging loops of many ropes caught on this sagebrush as the boys ran in frantic haste.

Sitting on our horses we saw these unfortunate ones brought up short and then be forced to waste precious time getting the loop off the sagebrush. Someone roped one of the wagon work horses and the cook ran up and claimed it, time was lost in a short but hectic argument over it, two loops were thrown on one horse, more argument, but finally all were fitted out with something, anything to ride.

In the midst of all the excitement one young chap ran up to me and said, "What are you going to do with the wagon?" "You may have it,"

I answered. "Oh, Hell," he replied, and then we were off on another wild race.

Out of the path of the storm we stopped and watched as it passed on the opposite side of the creek from our wagon. It worked havoc with the tall trees that lined the creek bank as it went bouncing along like a rubber ball, missing some places, dipping down and wreaking its fury in another. A little farther on it rose up and left the earth and was seen no more.

Again we were fortunate in that no one was hurt and the remainder of the evening was spent in going over the funny incidents of our scare. For old time cowpunchers never let slip an opportunity to "josh" the other fellow and now anyone whose actions had been at all ludicrous were told about it with a few frills added for good measure. All of which was taken in good humor, for all frankly acknowledged their fright.

It was this ability to give and take in the fun around the campfire that broke the monotony of our work and made for the comradeship that was a part of our lives; comradeship that has been felt all down through the changing years.

Chapter XXII

Proving Up the GG

As I have here-to-fore mentioned, the years of the early 1890's were a time of prosperity and opportunity for the cattlemen remaining in No Man's Land. In order to illustrate this I must tell of some other of our GG Ranch activities of those years.

The summer of 1891, Mr. Taintor bought 1,000 head of ♡ (Heart) Jinglebob cows, each cow with a steer calf by her side. Wm. Roberts, later more familiarly know as "XI Roberts," had recently purchased the XI Ranch on the Cimarron River with all the cattle on it. The Heart Jinglebob herd was one he had brought up from the old Chisum Ranch in New Mexico on moving up here. Now this herd was offered for sale. The purchase price paid by Mr. Taintor, if my memory serves me right, was $17.50 per head for the cows, with calves thrown in. The herd was received and branded at the XI Ranch and then taken to our home range. The calves, sold the following year on a rising market, paid for the cows.

The years of 1891 and 1892, were the years of greatest activity and achievement of all my years on the ranch. For besides all the routine work of spring and fall round-ups, line riding, branding, and cattle shipments to Kansas City markets, a building program was begun the fall of 1891. It was decided to now build stables and corrals as well as to remodel the Ranch House.

All the work was being done with the native stone. Forces of extra men were employed and all hands set to work getting out rock from the nearby hillsides. The old time cowboys of the early 1880's were hired and expected to do very little work except riding, some of them would do nothing else, but now the times were changing and, like or not, manual labor also was required of the men at all times.

The work of getting our rock proceeded rapidly, so when it was time for the October round-up for late beef shipment part of the outfit was sent away to the Indian Territory, below Camp Supply. The others remained at the ranch doing line riding and helping the stonemasons on the new buildings. A work that took all the winter to complete, but when finished this ranch was one of the best equipped in the country and that combined with the natural beauty of its surroundings made it one of the show places of the country at that time and for a number of years there after.

We had always taken pride in keeping the place up and making it attractive and home-like, from the very first it had been noted for its hospitality. My boyhood home had become but a pleasant memory. Now, after years of work on the ranch, the greater part of the time as its foreman, to me this place was home, for I had had an active part in the making of it.

We had a large built-in bookcase made when the ranch house was first built. It's shelves were filled with the best works of standard authors; Milton, Byron, Shakespeare, Poe, Dickens and many others. There were books of travel, essays, autobiographies, all in good bindings and this small, but excellent library was owned jointly by Mr. Taintor and myself.

Later when a Post Office was established nearby, we subscribed for many newspapers and magazines so there was always a plentiful supply of good reading matter.

On our return from beef shipment in Englewood that fall and while busily helping with the building operations and thinking that the cattle work except line riding was over for the winter, there came to the ranch a man from eastern Kansas. He wanted to buy a hundred head of two and three year old steers for his feedlots.

Taintor agreed to furnish them, so once again we started out. It was now November and we left the ranch at daylight the following morning. A snowstorm was on however, in our work, business came first and the weather was a secondary consideration.

That evening we made camp at Coon Creek, south of the Beaver River, at the home of my good friend, Henry Drum. Several inches of snow had fallen when I rode in there ahead of the wagon. Mr. Drum had his yoke of oxen hitched to the big sled and offered to haul in a load of fodder for our horses, an offer I gladly accepted, at the same time offering my assistance in getting the load, with the intention of having a little fun, for we were always playing practical jokes on each other whenever opportunity offered.

Just the summer before, while on roundup on the Beaver River, two chuck wagons were camped side by side and everyone was around the chuck tables getting their tin cups filled with coffee and plates of food. I was at one wagon, Mr. Drum at the other. He was a great talker and while busily engaged in the telling of a story he turned momentarily, so

Part 2 • Chapter 22

his back was to me, and in that moment I quickly poured a large helping of salt into his cup of coffee. His story finished he turned back to his food, put a heaping spoon full of sugar in the coffee, stirred it vigorously, and then took a good big swallow of the mess. Without a word, he dashed the cup down and made for me, and I ran

Close by was a house with a small lean-to, open in front and facing the wagons. Drum was close on my heels, but there was a footboard broken out near the bottom of this lean-to, and I being small, I wiggled through this opening. Mr. Drum of more portly build tried to follow and got stuck in the narrow apertures.

It was a situation too tempting to forgo, I seized a board and running back into the lean-to applied the paddle with gusto, while the whole outfit whooped and laughed and applauded.

When he finally broke his way out we went back and finished our interrupted meal. Everyone, even the victim, was in high good humor. Little incidents like that were all in the day's work and play.

So on this snowy evening, Mr. Drum and I started for the field with the plodding ox team, talking busily as we went along. His little six-year-old daughter had begged to go along so we had her with us. While loading the fodder at the field, I walked past the oxen and quietly dropped a small portion of "hokey-pokey" on each one's rump. When the oxen began to "perform" their astonished owner tried in vain to stop them, while I caught up the little girl and got her out of the way of the mix-up that ensured, for "Holstein" and his mate went around and around. Mr. Drum, trying vainly to check them, finally slipped and fell down, sled and all going over him.

When things quieted down, no harm done, we loaded the sled and went home. The horses were fed, cook had supper ready and when it was over we all went to the house to spend the evening. We were well entertained by stories of the Civil War, violin music and bugle calls for our host had been a bugler in the Union Army.

Probably it was the warmth of the setting room that caused the gas to expand in the bottle of hokey-pokey I carried in my pocket, for the cork became dislodged, releasing a very unpleasant odor in the house. An odor the boys recognized at once, and they were filled with glee when I arose and went outside, where I got rid of my bottle by placing it on the low roof of the part dugout house. On leaving later on I forgot to retrieve my bottle and so it was found by my friend and then "all things were made clear" to him, but he was far from being indignant over the trick for he was one who liked to play jokes on others and could take one on himself when his turn came.

It was several years before I again saw him. We happened to meet in town. The minute he saw me he shouted, "D___ you, you hokey-pokeyed my oxen." And we both burst out laughing. Then, in his own jovial way he recounted the whole affair to the assembled crowd. He was one who could meet the cowboys on their own footing, could join whole-heartedly in their pranks and jokes, in fact recognized them as folks just like himself.

Some others I knew attributed to us most of the vices and few of the virtues of ordinary mankind. As, for instance, the "old darky" who walked into our camp one morning on the Beaver River, he had a claim somewhere back in the hills, and was out looking for his horses that had strayed.

We had just finished breakfast and were ready to start on the round-up circle. He was invited to have some breakfast and wait at the wagon while the cattle were being brought in to the round-up that was to be held nearby. Then, if any of us found his horses while making the circle, we promised they should be brought in with the cattle. He looked ill at ease and very doubtful, but ate his meal and awaited development.

It was not long until two of the boys came in from the south with a bunch of cattle bringing with them the strayed horses also. They were soon caught, led to the wagon and turned over to their delighted and incredulous owner. He was profuse in his thanks and bowed low, cap in hand. As he prepared to leave he said, "Boys, I shore is surprised for I been hearin' all time how you boys is just a no-count bunch a thieves and now just see how fine you treat me." And then in a final burst of gratitude, he added, "I think the one's being tellin' me that is the thieves."

I have only one other outstanding memory that winter of 1891-92, which was my last one at the G G Ranch and that was a trivial accident that came close to closing my career very suddenly.

Bringing in a beef for slaughter one afternoon, as we neared the house and before entering the timber, I made a throw with my lariat. I missed the animal and the dragging loop caught on a picket pin left in the ground where a horse had been picketed. My horse was on the run and when we were brought up short by the sudden stop, in some manner that I have never been able to figure out, that rope took a half hitch around my neck and in the next instant I went sailing through the air and landed on the hard ground with a thud.

Half dazed I staggered to the house and was met by the cook who had seen the whole performance. Looking me over to see what damage was done he saw the great red welt all around my neck where the rope had burned it and seizing the turpentine bottle he doused the burnt place generously. Burn, Ye gads, I thought I was on fire.

I was terribly sore for a long time, and the burnt circle peeled off in a thick layer later on. But the only reason I escaped a broken neck was that the picket pin broke off just above the ground, for it was a wooden one, though we ordinarily used iron picket pins. By such narrow margins we sometimes escape sudden death.

Fish pond on the Taintor ranch east of the GG Headquarters.

Jack Bateman Claim, GG ranch on the Cimarron, 1892, Gerd Meierdierks in the foreground right.

Chapter XXIII

The Hash-Knife Herd

The summer of 1892 was a busy one for me. Besides my work at the ranch, I was supervising the building of a home of my own, for I was intending to be married that fall.

Mr. Taintor was in Kansas City for several months that spring. Returning from an early round-up trip, I found a letter from him to inform me that he had purchased two thousand head of Arizona cows and that at a certain date I was to be at Dodge City to receive them. He had made arrangements to pasture them for a part of the summer not far out from Dodge City. He was on the way to Arizona with five experienced shippers from Kansas City to load these cattle and care for them on the way in.

When I saw them I knew why they needed expert supervision on the way. They were of the Hash-knife brand, an old long established ranch in Arizona, their nearest shipping point was Holbrook.

This part of Arizona had experienced five years of severe drought. Cattle had died of starvation by the thousands in that drought stricken region and Mr. Taintor had bought these cows at $5.00 per head.

When the first consignment of twelve hundred head were unloaded at Dodge City I was undecided as to whether they had been bought cheap or high. I thought I had seen poor cattle before in my time, especially during that blizzard winter, when they died by the tens of thousands. But I did not know that cows could be so thin and bony as these were and still be able to stand up and walk. Of the twelve hundred head loaded at Holbrook one hundred head had died on the way.

The expert shippers from Kansas City were a blue bunch when they got in. They had had plenty of work and trouble in route, especially when coming across the mountains. They had prided themselves on their record of never having lost any cattle in their many large shipments and now to have lost a hundred head, it was terrible.

It was about the 5th of May, but the grass on the Arkansas River bottoms was fine, a heavy mat of old grass and through and underneath it was plenty of good green grass. We turned the bunch out to graze, just south of the stockyards.

Even in their weakened condition many of these cows came out of the cars strictly "on the prod." The old time range cows knew what it was to fight for existence and some of them developed a perpetual belligerence that only death could quench. But these cows were not old, some of them three and four years old, and they had never eaten real grass such as they were now turned out to.

They had lived on weeds, cactus, brush and such provider as a rainless desert afforded.

We did not attempt to move them farther that day. We just watched them as they ate and rested, scarcely moving out of their tracks, except to walk the few steps to the river where they drank long and deep.

Next day we crossed the Arkansas River with them, a number of the weakest ones had to be helped out, although the river was low and no quick sand.

After seeing them safely across the river, I started for home, leaving Bob Wright's cowboys to take them on to the pasture about twelve miles beyond. It took the boys two days to get them there.

Two of the shippers now returned to Kansas City, but the other three returned immediately to Holbrook for the last shipment and these were brought in with a loss of but fifty head.

About July 1st we went back to Dodge City with chuck wagon and full outfit to bring home the Hash-knife herd as well as a carload of saddle horses shipped in from Colorado.

The cows left unmolested on good pasture for two months were now fat, sleek and wild. No one would have recognized them as the pitifully weak, emaciated bunch that had arrived only two months ago.

That trip home with the Hash-knife herd was a hectic one for the entire outfit. The cattle were wild and the Colorado saddle horses even wilder, although the seller had guaranteed them all as well broken saddle horses.

Well broken, hah! They had most of us well broken before we got well acquainted with them. Added to all this, we soon found that we had a crazy man on our hands.

Just before leaving town, Mr. Taintor, who was not coming out with us, came to me and said that a friend to both of us had asked a favor. It was that we take out with us a man to stay at the ranch for an indefinite time in the hope that the trip and absence from town would help to overcome the unfortunate condition he was in from weeks of heavy drinking.

Part 2 • Chapter 23

He had already suffered an attack of delirium demons and was on the verge of another. A talented man of fine education, he had ruined his life through the habit that now held him a victim. But he had friends who were still trying to save him from utter destruction.

My boss left it up to me to say whether or not he should be sent out with us and I carelessly consented to take him along. For convenience in referring to him, I shall call him Mr. James, although that, of course, was not his name.

The first night out from the pasture we camped high up on the Mulberry and there was trouble from the start. Just when the cattle were bedded down for the night and horses were brought up for the night shift, one of the horses when roped jerked loose from the man who had roped him. Mr. James, with well meaning, but nervous energy seized the rope and hung on despite his being thrown off his feet and rolled over and over and this so frightened the horse that he stampeded into the midst of the cattle and they were up and away also.

It was a bright moonlight night and we kept them together but they were nervous and restless all the night through, so none of us got any sleep.

Next morning, when the horse herd was brought in and we were catching our mounts for the day, about a dozen of the Colorado horses bolted out from the side, out ran the horse wrangler and were soon out of sight. We naturally supposed they would run a few miles and then stop to graze. When Jack Bateman started on their trail sometime later we expected him to bring in the runaways in a short time, but nothing like that happened. In fact, it was some days before we saw Jack again. We were home with the cattle when he returned with just a part of the bunch. These horses after leaving us ran for some miles in the wrong direction and then seeming to get their bearings they headed straight for the northwest, toward Colorado and the home range. They must have found many obstacles in their way for there were many fenced pastures and fields, but with the peculiar homing instinct of all wild things, they persisted on their way. They had become scattered, however, so when Jack found the first bunch he brought them home. He then went back again with Wiley Rhodes to help him find the others.

They were found scattered about in different pastures in the vicinity of Cimarron, Kansas. At one place they found the meanest and wildest horse of the entire lot and on his back was a small boy not yet fully recovered from a broken leg. The boys were horrified, for we had all been afraid of that horse and here was this little boy with a quilt strapped on for a saddle, a "blind" bridle on the vicious beast, enjoying a nice quiet ride. The "Lion" was being handled by the "Lamb."

Every one of these horses pitched like demons when we rode them. One of them had thrown a man clear over the chuck wagon and another had cleared the cook's table, he was pitched so high. No wonder the boys were goggle-eyed when they saw this urchin sitting so calmly on what to us had been plain dynamite in the disguise of a horse.

With the horses assembled and started for home once more the usual trouble started all over again for they would stampede. Finally after a straight twelve-mile run they were corralled in the stockyards at Fowler, Kansas. Tired and disgusted the boys bought stout straps and small lengths of chain to attach to a front leg of each of the outlaws and so they were brought in to the ranch somewhat subdued for the time, but they never became gentle or trustworthy.

Now, to return to our Hash-knife herd, the day following our sleepless night they were tired enough to be quiet and rest. We were three or four miles from the head of John's Creek in Clark County. Mr. James had not given us any real trouble, but was extremely restless and nervous when we started out next morning.

I told the cook to drive ahead, find a good watering place on John's Creek, fill his water keg and drive farther on to make camp so the cattle would not be disturbed while on the water. He followed instructions and had unharnessed his team and the horse wrangler had driven them in to the horse herd when we heard a commotion at the wagon. Returning we found Mr. James chasing the cook around and around the wagon and threatening him with an open knife in his hand.

Tragedy was averted by hastily producing a stiff drink of whiskey for him, but the resultant quietness was only temporary. Soon after dinner was over I saw the cattle running wildly in fear from some object. Riding hastily to the scene, I discovered Mr. James walking down the middle of a deep sand draw in the hot July sunshine and he was walking fast. Quickly I caught up with him but he looked at me with vacant eyes. "God," he said, "the pavements in this town are hot! But I have an appointment, it is important." And he hurried on.

Riding along I talked quietly and finally persuaded him to go with me to the wagon where he rested quietly for a time. I stayed close by him until bedtime when we put up a small teepee tent for him to sleep in while the others rolled up in their blankets nearby.

He retired but not to sleep for I heard him talking to himself all the while, so I went in and sat down by him thinking I could quiet him, immediately he said he meant to get up and dress. I said, "Oh no, don't do that, everyone else is in bed." Quick as a flash he reached for a shoe

Part 2 • Chapter 23

and aimed to bring the heel of it down on my head but I jerked back and the blow fell on my chest. At my sharp exclamation one of the boys thrust his face inside the tent flap and got a black eye from a well-aimed kick in the face that sent him sprawling backward. We then had a regular "rough house" for it took the combined efforts of us all to hold and subdue him.

I told the boys then that a crazy man and a bunch of nearly two thousand wild Arizona cattle were entirely too much for one outfit to handle and that I would find some one to take him back to town.

About two miles back from our camp was a house and two strong, sturdy young chaps lived there. These I hired with team and wagon and about daylight they started with their charge.

To make sure that he be returned safely to his friends and not to be too roughly handled in case he became troublesome again, I sent one of our men along. However, his only break on the way in was when his guard, tired and cramped from riding got out to walk behind the wagon that was minus the end gate. Mr. James had refused to ride on the spring-seat but persisted in sitting on the wagon floor with feet swinging over the edge. Now as the unsuspecting guard plodded along close behind, he got in a swift and effective kick in the big fellow's face that sent him sprawling in the road and he came up fighting mad. Perhaps it was well that I had sent a man along to see that our charge was not mistreated.

Relieved of this burden, but short two men, we finished our trip home without further trouble. The cows were taken to the Beaver Range and turned loose where they continued to put on flesh until late fall. At that time all calves large enough for veal were shipped to market. They brought $8.00 per head, so the Hash-knife deal turned out to be a very good one after all.

When I left the ranch in November there were ten thousand cattle in the G G Brand. The business continued in a flourishing condition until Mr. Taintor sold out some years later.

Then the old ranch, so well kept up during all the years of his ownership, soon was neglected and fell into sad disrepair. Floods swept the pretty creek bed and worked great havoc. All of the undergrowth and most of the timber is gone, the buildings falling into ruins. Little now is left of what was once such an attractive spot. Only some of the great cottonwoods, scarred and hoary with age, still stand sentinel like watching the ravages of time.

Healy Brothers Outfit on the Beaver, ready for roundup circle.

Chapter XXIV

The Cimarron River

It is always interesting to reflect on how, when, and by whom the main streams of a country have been named. A great many of these facts have been lost in antiquity.

I have mentioned that Cimarron is a Spanish word of several different applications and meanings. The one that applies to the river, no doubt, is that of "Wanderer."

When the first survey for this historic Santa Fe Trail was made by the U.S. Government, something over a hundred years ago, the dead surveyor, J.C. Brown, in beautifully fine clear script marked this stream on the map as "Semaron" and this spelling perhaps came nearer to the true Spanish pronunciation than does our "Cimarron."

The greater part of my life has been spent near this river. When I first knew this stream it was much narrower than it now is and it flowed between grassy banks. In many places the rushes grew near the waters edge tall enough to hide a horse and this was evidence that for many years no floods of any great violence had disturbed the meanderings of the river.

In consequence, almost all of the first settlers along the stream built their little sod homes very close to the river and dug shallow wells for household use. They depended entirely on the stream for watering their livestock.

The Indians, however, had a tradition that this river at flood stage would stretch from "hill to hill" and that sight was a terrifying one. It was fortunate indeed that a recurrence of this did not take place in the white man's time until all of these squatter homes had been long abandoned and where substantial ranch homes were built farther back from the river.

The Cimarron is a historic stream, in normal times a beautiful and useful one. Its broad meadows have supported thousands upon thousands of cattle in luxury and plenty. The hills on either side furnishing shelter and rich grazing as well as beauty of scenery, for these hills are almost constantly bathed in a soft blue haze.

At evening the view from one of the high points of these hills is a most entrancing one. One can see miles and miles up and down the peaceful valley, with the river, like a ribbon of silver, winding through and at times hidden by a bend then again gleaming brightly as it emerges into view. It is a picture that an artist would find impossible to convey to canvas.

This is the river in a mellow chastened mood. But "Old Man River" at high flood is something very different. Some of us lived here to see when the cloud burst on the river's upper range and torrential rain fell along its entire course. A wall of water swept everything before it and the river came thundering and roaring down the peaceful valley and in many places it extended literally from hill to hill (1914.)

The roar of the angry water was deafening close by and could be heard for miles back from the river. It left a trail of ruin and desolation that twenty years of time has failed to entirely obliterate.

Since then the river no longer flows in a narrow bed between grassy banks but its bed has been a wide shallow sandy one with many changes of channel and much quick sand, a menace to man and beast. Had this flood occurred while these first settlers had their homes so close to the river's bank many whole families would have been swept away.

It was on the Cimarron that I located a homestead and the house was placed on the north side of the river near the place where the old Healy Trail crossed the stream.

My wife, who was Miss Carrie Schmoker, and I were married November 17, 1892, at the Schmoker Ranch home on Crooked Creek. A few days later we came to the home we had ready and waiting and here we have lived continuously now for more than forty-three years.

My work on the range was over for soon all of the smaller ranch holdings were placed under fence, the bottom lands furnishing abundant hay for winter feed. Orchards and groves were set out and cultivated. We had small fields of corn and forage crops for now the seasons were, for a time, more productive.

A new lot of settlers came in with the opening of the "Strip" for homesteading and many of these remained permanently.

We were seldom without visitors in our home during those years. Living so near the river crossing, a much used one at that time, our home was a convenient overnight stopping place for all the herds driven from the south to Meade or Englewood for beef shipment. We

Part 2 • Chapter 24

were always in close touch with the old cowboy friends remaining on the ranches and they were ever-welcome visitors.

The settlers too, with loads of grain, in route to town or returning with supplies, invariably stopped for the query "How is the river today?" The river was ever a factor to be reckoned with.

In summer it was high water or quick sands following a rise. And in winter the ice that sometimes for weeks at a time made the crossing a hazardous one.

Always in time of trouble or accident we were called on for assistance. I can still see in my memory the heavily loaded wagons that broke through the ice, or as often happened, a double tree would snap. Then a shivering man would shed boots and socks and with trousers rolled high, wade into the icy stream to repair the damage. Meanwhile the load settled in the sand and stuck, so we sometimes worked for hours in freezing weather and icy water to extricate the outfit, the load carried out piece by piece on men's backs.

In summer the quick sands were treacherous and sudden rises in the river came all unlooked for. On one such occasion a neighbor drove into the stream with a load of kaffir corn on his wagon. In mid-stream the load stuck for there was a slight rise. He unhitched the team and came to the house for help. Returning some thirty minutes later the river was up, definitely so, for the load of grain, wagon bed and all had lifted and floated away. The wagon bed was never found but some days later we recovered the running gears.

Once a string of wagons loaded high with bales of broomcorn reached the crossing just after dark. They were from Beaver and bound for the railroad at Englewood. The lead team drove in. As often happened, a deep hole had washed out. In the dark this hole was struck by the wheels on the down stream side, the top-heavy load lurched to one side and turned over. The driver of this load had his wife along and both were thrown far out in the river. They as well as the baled broomcorn were well soaked. The lady was brought to the house and dry clothing furnished her. Everything was several sizes too small, to be sure, but this discomfort was passed over lightly and only furnished considerable amusement for all.

It was a bright moonlight night so the men managed to recover the broomcorn bales without much trouble. But the buyer at Englewood must have wondered a little about the weight of that load.

Some two and one-half years after our marriage our first child was born, April 17, 1895, a boy named Wallace after his father. I had always been known by the nickname of "Doc," so now my wife said the name should be given to our boy and that there would be no nickname.

Something over two years later (July 13, 1897,) a dark-eyed little daughter, Esther, our only daughter, brought more cheer to the home. Then seven and a half years later (December 2, 1904,) our last child, another boy, named Fred was born.

The years sped on. All too soon, the two older children passed grade school and had to leave home for high school. At that time the youngest boy was just entering grade school, a fortunate arrangement for us, because for the parents when the children leave for school and college it usually is the beginning of the end of the family life in so far as the children in the home are concerned.

The little white schoolhouse was on the opposite side of the river so all through grade school years the three children made their daily pilgrimages across the river at all seasons and weather and river conditions.

"Old Mike" a fat and faithful boy pony was their most dependable means of transportation. Mike soon learned to meet and cope with all the vagaries of the river. In some mysterious way he seemed to scent quick sand and could pick out a safe way.

When ice began to crack under his weight he never got excited and slid or scrambled about as some horses do. Instead he stood still and teetered until he broke through, then calmly climbed upon the ice again to repeat the performance until the farther bank was reached.

Truly, the horse and the dog are man's best friends and most faithful servants.

The fall of 1911, work was begun for the grade of the new railroad across the flat between the Beaver River and Cimarron River some six miles to the south of us. Grade camps were established and all supplies for these had to be freighted from Englewood. It was a terribly cold and snowy winter. Teams were on the road constantly for beside the grade camps that had to be kept in food, fuel and grain for the horses and mules, a great amount of cotton cake had to be brought out for the cattle. For many, many weeks snow covered the ground.

It was claimed that the snowfall that season totaled sixty-five inches. That winter our place was taxed to capacity for room and beds for the freighters and some sort of shelter for their teams. To give some idea of the amount of food cooked and served by my wife during that winter it will suffice to say that we killed five large beeves to supply the meat for that time.

Part 2 • Chapter 24

No man was ever charged for bed or board, and I do not make this assertion in a spirit of boasting, but merely to illustrate the spirit of hospitality that was a part of the code of the old time West.

Wallace Anshutz, Carrie Schmoker Anshutz, Esther Anshutz and Fred Anshutz.

Doc and Carrie Schmoker Anshutz ranch home on the Cimarron River.

Chapter XXV

The Flood

It was the spring of 1914 when the big flood came. The two older children were away at school. Fred was just a little fellow then. We had telephones in our homes here on the river at that time, but these gave us no warning of the awful thing that was in store for us.

The night of May 3rd there was a severe storm, rain fell in torrents for hours. We could hear the river roar for a time and then become quiet. We were up and out early next morning to see what damage had been done by the night's storm.

A gray pall seemed to lie over the valley so the visibility was poor. In the dim light we saw a bunch of horses in a bend of the river to the west throw up their heads in sudden alarm and prancing wildly. The next instant a rider at full speed was chasing them out to the hills, and then we knew, the River.

Our house and all our buildings had been built back from the river against the first low line of hills. Several times in previous heavy rains the water had backed up into the yard, once or twice it even encircled the house, a sluggish back water just a few inches in depth that soon subsided and never had given us cause for alarm.

To the west and south of the house we had set out a grove of trees to break off the prevailing winds and these had made a wondrous growth. Farther to the west was a fine peach orchard completely surrounded by tall cottonwoods for a windbreak. To the east of the house was the garden and beyond that a fine field of young wheat.

We were all working hurriedly to pick things up out of the way of the flood when the horseman, having run the horses to the hill came riding up. It was my wife's brother, Will Schmoker, who lived a mile up the river from us. "The River," he cried, "is going to be higher than we have ever seen it."

During my boyhood in Ohio I had experienced several floods when our home was damaged by high waters of the Ohio River. So I felt no fear, but called to the young boy, who was working for me that spring, to come with me to the house to get things up out of the way of the water that might seep in. At the same time telling the others to get out onto high ground or come to the house.

None of us, however, reckoned with the speed and volume and terrific force of the flood that was now upon us. My wife and Fred were in the chicken house gathering up a lot of young chicks to carry out to safety. Her brother, Will, had tied his horse back out of the way and a minute later appeared at the door of the hen house saying, "Come, we must get out of here." At the same time he grasped the arm of his little nephew.

A thin stream of muddy brown lapped the feet of the three as they stepped outside. To the north just against the henhouse was a wire fence that they must get through. Will and Fred went around to the west of the henhouse to meet the flood. My wife turned to the east, away from it, rolled under the fence and ran the few feet toward higher ground when the first wave struck her and swept her off her feet. She felt herself being rolled over and over. With an effort she grasped some sage, growing along the edge of the hillside, dug her heels in the ground to help hold on and then slowly rolling over and over in the muddy water, pulling herself from one bunch of sage to another a little higher up where she managed to reach dry ground. She stood up and saw with horror, Will and Fred just behind the hen house in swirling water waist deep, clinging to a fence post.

"Get up on the roof," she called to them, shouting to be heard above the angry roar although she stood scarcely twenty feet away. "I am afraid to let go," Will shouted back. But the water was rising higher, with a great effort, he lifted the small boy onto the low, nearly flat roof and then by means of tearing off a window screen and getting a footing in the window he pulled himself up.

Safe at least for the time being, but how long would that frail structure withstand the pounding force of that flood. Only the fact that it was held by thick cement flooring kept it from being swept away at once.

Meanwhile inside the house strange things were happening. We got inside just as the first wave of water struck against the west side of the house. We closed doors and windows, but the force of the rushing flood soon broke some of the windowpanes, for the water was up to the middle of the sashes.

The outer door burst open and some things washed out and away before we could get the door nailed shut again. Soon the furniture began to topple, I heard a noise behind me. The cook stove had opened and fallen forward completely upside down. The heating stove in the front room did likewise. The heavy piano fell forward on its face. I picked up what I thought was a necktie but it was a striped snake. The water was up to our waists and when we got too tired of wading in it we could go upstairs where it was dry. But mostly we stayed down where we could better see what was going on outside.

Those two figures perched forlornly on the top of that hen house worried me not a little. The water was getting higher. Then I saw my wife get the saddle horse and at a run she was off to the east for help, I knew.

The river was a mile wide just below our place and the freight that it carried. Great stacks of hay, many of them were dipping and bobbing as they were carried swiftly along just about half submerged. There were buildings of all shapes and sizes, great rolls of wire and fence posts, trees, livestock and some objects that could not be classified. But what claimed my wife's attention the longest, in spite of her pace, was out in midstream - an automobile!

It did not seem possible, but she was sure of it. Weeks later we knew she was right for some miles farther down the river it washed close in shore and was salvaged. Report was that it had come from somewhere in Colorado at the start of the cloudburst.

By the river road, the distance to the Spurgeon home was about a mile. But this road was now impassable in all the draws leading to the river. Water had backed up so it would swim a horse, so she was forced to go farther back through the rough hills and canyons.

After arriving at her destination and making her errand known, two men were quickly on their way, not stopping to go around. Their horses swam all the intervening draws.

Stopping at the house only long enough for a change of dry clothing, she was back on her way home. Riding more slowly now, for she felt sure that the rescue from the hen house roof would be easily accomplished by a man on horseback. But not so, they were waiting for the horse she was riding, a good steady roping animal. Slowly and steadily he made his way through water only knee deep but unbelievably swift. To the back of the hen house, the small boy was tossed to the man in the saddle, caught and brought out to his waiting mother. Riding in once more, a rope was tossed to the man who fastened it around his waist and then he slid down, feet first, into the water. As if he had been a feather he was swept off his feet and carried the length of the rope. The resultant pull on the other end that was fastened around the saddle horn made the horse lose his footing and he too was carried along a short way. But now he braced himself, regained firm footing and slowly, very slowly, pulled the recumbent man out onto dry ground.

We all felt better then, the group outside came as close as possible to the house. I broke an upper pane from a north window to tell them not to attempt any rescue for us for we were safe enough.

All that long day the water beat and rushed and roared against the house but it stood firm. It was protected somewhat by the surrounding grove that caught much of the debris. Not even the tops of the posts in the garden fence were visible.

That evening after a full twelve hours, the flood began to subside somewhat and we were able to leave the house. For some days longer the river was a quarter to a half-mile in width, an ugly, angry, menacing thing.

Our place was a picture of ruin. The work of years destroyed. The peach orchard and cottonwoods on the west, as well as most of the grove on the south had been uprooted and carried away. The soil and that of the wheat field washed away to a depth of four feet down to pure river sand. Our rich hay meadows were in the same condition. All river sand and great holes left, filled with water. Machinery nearly all ruined. Fence posts and wire all gone. The house furnishings were ruined, the sitting room floor broken down by the weight of mud and sand that the flood left.

Kind neighbors from the north came and helped us clean up some of the wreckage. Our neighbors on the south could only look across the sudden sandy waste and wonder how we had fared for our telephone lines were all gone. For several weeks no one wanted to attempt to cross the still swollen stream.

One day I decided to try shoving the loose horses across in order to settle the quick sand and so effect a crossing. Much against their will we forced them in. When nearing the south bank they suddenly dropped completely out of sight. One by one heads reappeared and they wallowed and rolled and fought in the treacherous quick sands.

The whole family had come out to watch and while the horses made their gallant fight, little Fred let out a heart breaking wail, "Oh, Rose! My pretty Rose, she will be drowned!"

Rose was his pony, a bright bay mare, and indeed for a time it seemed that all of them might drown. But they turned and made their way back to us on the north side. And we tried no more experiments of that kind.

When we finally did cross the river it was on foot when the water was down to normal depth and we carried a stout "prod" to feel the footing ahead.

Part 2 • Chapter 25

Ours was a discouraging outlook that spring and yet we were more fortunate than some. At Point of Rocks lived a family who had experiences somewhat similar to ours but with more disastrous results. Out of their family of three little daughters, two were drowned. Their ranch house was made of adobe. It collapsed under the fierce onslaught of the flood.

As soon as possible now we moved everything to a new home site a quarter of a mile back from the original one and on high ground. There followed the long hard task of again building up a permanent home.

It was not long until the chill shadow of the World War appeared on our horizon. Wallace, in his second year at Kansas University, left there at mid-year to enlist. By January 1st he was in training camp in Florida and about April 1st, as a member of the American Expeditionary Force, was onboard a Transport bound for France.

There followed months of anxious waiting. The infrequent letters from "Somewhere in France," carefully censored, brought us little of what we were eager to know.

Here at home we worked harder than ever before in our lives I am sure. When, after twenty-two months of service, he was mustered out. He returned home, his boyish enthusiasms somewhat dimmed. Thoroughly disillusioned as to war and any ultimate good to be derived from it.

His travels, after enlistment, in the United States alone had carried him across all the southern states, almost the length of the Atlantic seaboard states and back through the north central and mid-western states. He now announced that of them all, the mid-west and our own plains country was his choice of a place to make a home.

Running true to form the West had put its "'brand" on him too. Or perhaps the "brand" was an inherited one. At any rate to him of late years has been passed much of the burden and most of the vicissitudes pertaining to the farm and ranch. The other children are "city dwellers," but they too bear marks of "the brand" as attested by their frequent visits to the old place.

So in recent years, with more leisure, together my wife and I have revisited most of the scenes of my early cowboy days. In a trusty little Ford, we rambled along close to the foot of old Greenhorn Mountain, where high up on its wooded sides, I had helped get out railroad ties and then watched as they went leaping and hurtling down the chute.

By a safe and splendid highway, we crossed the mountain pass. There over a narrow, rough, and dangerous road I once drove a six-mule freight team with jerk line and bells on the harness to warn oncoming teams to wait at the passing places.

We have followed the wandering of the Old Cimarron River almost from its source in Colorado through New Mexico. It runs most of the way through Kansas and a part of Oklahoma where the water, sand freighted is always cloudy in appearance. At Keystone, some eighteen miles west of Tulsa, it flows into the Arkansas River and its waters there a deep brick red hue due to the red soil of that part of Oklahoma.

Historic Old River. Generations of Red Men have traversed its restless length. Coronado and his band of thirty intrepid horsemen probably camped on its banks to rest and partake of the buffalo meat. Most certainly they forded the stream as disappointed in their dream of finding the Seven Cities filled with gold, the fabled cities of Cibola, and made their way back across what is now the Oklahoma Panhandle.

Thrifty, adventurous French traders knew its ways in that long ago time. Early explorers, buffalo hunters, soldiers, trails men, freighters, and cowboys, all drank of its waters. Like links on a chain each made their contribution to the development of the west. An age-old river, what mysteries and tragedies may lie concealed by its shifting and engulfing sands.

To the West and East, North and South, we have been over most of the routes I traveled on horse back in my round-up work. But in all of our wanderings, over the once familiar ground, there is scarcely any resemblance now to what it once was. Only a few landmarks left, for everywhere there has been change.

Of the old trails leading out from Dodge City, spreading like the fingers of a huge hand to the far places south and west, there is now no single sign left. They had their day and they served well. But everywhere now are splendid motor roads. On a recent morning in a modern motorcar we were driven to Dodge City in an hour and twenty minutes, a trip that fifty years ago took from two to three days. Fifty years hence, in a trim little family aircraft, people may make that distance in a few minutes. For the tempo of life is faster, ever faster.

Part 2 • Chapter 25

Doc Anshutz, D. H. Simpson, and Archey Keech, taken August 22, 1929, in Dodge City, Kansas. "After forty years."

Doc Anshutz (left) standing with an unknown man.

Conclusion

Almost sixty years of life here in the Plains Country has brought many changes. As we grow old we are apt to deplore many of these changes and yet in the main, change means progress and without progress we attain nothing.

But for the cowboys and cattlemen of the earliest and best years, those years held a charm that was felt by all. A something that tugged at the heart strings when, in later years, living again in memory of the old carefree days of the open range. The old cowpuncher always expressed regret for the days gone by, regret that the farmer with his plow had invaded this cowboy realm.

One and all of the old boys have voiced this time and again as with wistful eagerness they recounted tales of the old times and the old associates, always wishing they could have the old days back once more. But time goes relentlessly on and with it comes change. So I have seen this country change from the days of the buffalo to the years of rich attainment of the cattle business. Then through greed, the over stocking of all the open range, when nature with one giant gesture cut down the surplus with a great blizzard and at the same time impressed all with the obvious need of a different method.

Then came the settlers taking the country with a rush. Many of these settlers never intended permanent settlement, only the wish to use a homestead or preemption right. Then in the shortest time possible they sold out to get enough to move farther on. Thus for a time, it made a transient seem a permanent development.

These transients, however, made it possible for the permanent homemaker to acquire title to land and range. It enabling the smaller cattlemen to gain a foothold, so many small ranchers developed along the streams. Ranchers, who by combining roughage farming for winter-feeding for their stock, attained a very fair measure of success.

Then came the years of the World War and the high prices for wheat, tempting everyone. So once more the Plains Country was despoiled and robbed of its rich protection of grass. The soil of every sandy, hilly, Yucca studded pasture was laid stark and open by the plow. Again nature has taken a hand by sending a relentless drought and wind, which has always been a part of this country and probably always will be, and has shown the dwellers of the plains that a change is again imperative. We cannot go on as we have been. For the dust storms have worked havoc in the Plains Country.

The beauty, which in the olden days, was mainly in its rich luxuriance of native grasses, lovely wild flowers, shrubs and trees that grew in all the sheltered places in the rough country. The trees disappeared almost magically with the advent of the settlers. A few trees were set out to replace the ones destroyed. Fields were plowed where the soil should never have been disturbed. Pastures have been over-grazed until they are as barren of vegetation as the barren fields.

We have always had years of drought and these invariably bring much windy weather. While the land remained unbroken these high winds carried little dust, though in sandy places the sand blew to some extent, but did not carry far.

Some years ago my aged mother came out from Ohio to visit us. It was in the horse and buggy days. One particularly windy day we had been out driving and on reaching home some members of the family complained of the sand that had been blown in our faces and clothing. Mother laughed and said, "But your sand out here is such clean dirt, it shakes right out of one's clothes and no harm done."But the wind now carries a silt so fine that it penetrates wherever air does and it is far from being 'clean dirt'.

Out here in the so called "dust bowl" in the farming communities where almost all of the soil has been broken, many, many once prosperous and well improved homes have been almost ruined.

Fences and farm implements are buried under the dirt. It is piled up against some of the houses until the lower window sashes are buried in it. The outbuildings are half buried, the pastureland so covered with the soil blown in from the fields that one can scarcely distinguish between the two. Truly a desolate sight and the living conditions of these people have been almost intolerable.

When for days and weeks, almost without ceasing, the air has been so dust filled that the sun is obscured. There is never a breath of good clean air. The damage to the health of the people has been very great, not to mention that to property for that has been almost incalculable. In all our experiences, we old settlers have never before seen such dust storms as these.

Black Sunday, April 14, 1935, was the very worst of these storms. It came on about four o'clock in the evening after rather a fitful day and it was a sight to strike terror to the heart of any onlooker.

Part 2 • Conclusion

High in the air, carried ahead of the wind, were great inky masses of dirt. Clouds rolled over and over, over and over, and flying before these clouds, shrieking their terror were birds of all kinds and sizes. Many of these perished for they were found afterward lying dead all over the country.

As we stood outside watching this amazing sight, in one instant of time, utter darkness enveloped us. Darkness so complete and so impenetrable as no night in my experience has ever been.

Many persons were terrified almost beyond belief, some thought it was the end of everything earthly. For two hours this condition of dust and darkness prevailed. The homes were so filled with the dust that came through every crack and crevice that the lights burned dimly as if lit through a fog. After a couple of hours it lightened a little so that one might venture out. But for most of the night it raged on depositing its tons and tons of dirt in and over everything.

This was one of the "Northers" that for suddenness and intensity are always worse than the prevailing southwest winds for which the plains have always been noted. Often in midsummer these winds are hot, scorching hot even at night, long after the sun has gone down. But even these parching winds have never damaged the country as these dust storms have.

The whole face of the country has been changed by them. It is stark, naked, desolate looking where once it was richly clothed and protected even in the worst drought, and the air was pure, clean and sweat, where now it is just the opposite.

That these conditions have been brought about to a great extent by the people themselves under the mistaken idea that this was strictly a farming country does not make the situation any easier to bear. But the fact remains that great damage has been done to a once fair land and that something must be done to repair and restore it. Nature itself, if given a chance, will go a long way toward this accomplishment. But the people who continue to live here must give their help and serious consideration.

Even so the task is going to be a long hard one in order to win back the beauty and prosperity it once held. To the old timer it still seems that the old, once successful arrangement is best. That is of a small ranch located on a stream with range extending on either side and with enough farm operation to supply roughage for winter use.

The high flat dweller too must not depend on wheat alone, but must diversify his farming in order to stay with the country. I do not say that this is the solution. It may be that an entirely different way must be worked out. This is the problem that faces our children and grandchildren. It is to cope with these problems and the fast changing times that these young people are being given. The opportunity for education and mental development that was denied many of their forbears. The old folks were educated mainly in the University of Hard Knocks and often of bitter experience.

It is my belief that from time immemorial there must have been some sort of Gold Rule of Life, whereby people shaped their destinies in their upward climb from the crudest form of life to our present proud claim of "advanced civilization."

The "Code of the West" was founded on this Golden Rule. Traders, trappers, trailmen, hunters, and cowboys lived by this simple code that was to them the only law of the land. The Code that recognized the rights of each fellow man, and they lived it whole-heartedly, joyously through days and weeks and years.

The work and the hardships, the excitements and adventures, and just the simple things of nature by which they were constantly surrounded in their life in the open are the treasured memories of a bygone time shared by all of those of the pioneer days. After all it is the simple joys of life that are the riches of existence.

Rough and unlearned many of these men were. But they lived accordingly and adapted themselves to the changing conditions of the times as life flowed on. And that, no doubt, is just what their successors will continue to do. For after all, harsh nature is much the same as it was sixty or a hundred years ago. There will always be some who fail, some who violate the "Code of Life," but the rank and file move steadily on.

To these we leave this work of the future development of "Our West."

Index

A
A H outfit: 192
Adobe, Walla: 95, 127, 128, 130
Adobe Walls fight: 88, 89, 95
Adobe Walls Trail: 95, 109, 117
Alamosa: 119
Anadarko: 149, 166
Anchor D Ranch: 150, 177
Anderson, Bill: 181, 182
Anderson, George: 115, 141, 152, 177
Anthony: 149
Arkansas River: 13, 44, 56, 61, 86, 92, 95, 109, 115, 116, 120, 127, 128, 134, 137, 18, 157, 223, 224, 240
Artesian Valley: 103
Ashland, KS: 31, 95, 102, 103, 118, 121, 155

B
Bailey, Bailey Ranch:131, 139
Bar, A.H.:56
Bar, A.B.: 45, 59, 68
Bar Links Brand: 146
Barber County: 192
Bartlett, Jor: 181
Bateman, Jack: 180, 222, 225
Bear Creek: 118, 146
Beard, Jim: 215
Beattio Brothers: 127
Beaver City: 76, 96, 117, 135, 176, 197
Beaver Range: 235
Beaver River: 68, 116, 117, 126, 127, 130, 134, 148, 150, 152, 154, 155, 166, 177, 178, 186, 187, 190, 197, 204, 208, 209, 220, 228, 232

Beebee: 186
Beeson, Chalk: 88, 89
Bell, Ham:87, 88
Belle Meade: 19, 36, 67
Benton, Oklahoma: 187
Bents Old 9Fort: 120
Berends: 186
Beverly: 88, 96
Beverly Brothers: 103, 155, 197
Beverly, Judge: 115
Beverly Range: 177
Big Springs Ranch: 29
Biggars, Frank: 148
Blair, Peter: 21, 27, 41, 43, 44
Blossom, 29-31
Blue Moles: 103
Bluff Creek: 95, 102, 103, 107, 126, 155, 164, 177
Boland, Beck: 172
Boyer, Edward: 52, 53, 67
Braidwood: 186
Brown: 186
Brown, Frank: 88
Brown, Joseph C.: 92, 93, 229
Brown, Hoo Doo: 117, 148, 167
Brown, Wiltz: 209
Brushy Bill: 118
Buckner: 103
Budley, Ed: 177, 078
Buffalo Creek: 117, 179
Buffalo Jones: 90, 179
Buffalo River: 118
Buffalo, OK: 179
Bullard Outfit: 34, 128
bull-whacker: 97
Burton, Doc: 109, 110, 115

C
C X Outfit: 127
C.O.D. Ranch: 189
Caldwell, KS: 149, 164
Calvary Creek: 155
Camp Creek: 117, 215
Camp Supply:95, 56, 109, 117, 130, 146, 149, 150, 166, 167, 217

Campbell, Jim:88
Campbell, Johnny: 177
Canadian River: 44, 96, 127, 128, 134, 148, 190, 203, 210, 213
Cantonment: 149, 150, 164, 165
Carey, Nola: 117
Carey and Lane Road Ranch: 134
Carter, Frank: 38, 55
Castor Bean: 206
Catlin: 117
Cator, Bob & Jim: 126, 127
Cavalry Creek: 102, 177
Cherokee: 149
Cherokee Strip: 210
Cheyenne Bottoms: 115
Cheyenne Indians: 65, 107, 192
Chief "Stone Calf": 164, 165
Chief Buffalo: 165
Chiquita Creek: 126, 170
Chisum Ranch: 217
Choate: 116
Choteau, Dr.: 88
Christoffels: 67
Cimarron and Crooked Creek Cattle Company: 139
Cimarron Crossing: 117
Cimarron River: 33, 46, 53, 55, 56, 73, 84, 92, 93 ,95, 96, 101, 116-118, 127-131, 139, 140, 146, 150, 151, 155, 158, 167, 168, 173, 175, 177, 183, 186, 187, 190, 192, 195, 200, 202, 204, 210, 211, 217, 222, 229, 230, 232, 234, 240
Cimarron Territory: 208
Cimarron, Kansas: 75, 109, 225
Clark County, Kansas: 88, 94, 118, 146, 155, 171, 174, 185, 226
Clear Creek: 117, 126
Coffee: 19
Coffman, Charley: 118
Coldwater Country: 178
Coldwater Creek: 126, 150
Colgan: 19

247

Comanche County, Kansas: 155, 164
Comanche Pool: 147, 149, 168
Connell, Tom: 109
Connell Ranch: 110, 115
Conrad: 19
Coombe Brothers: 97
Coon Creek: 218
Cox: 117
Craig, Pete: 183
Crawford: 104, 117
Crooked Creek: 19, 23, 28, 33, 36, 44, 53, 55, 84, 101, 103, 108, 109, 116, 117, 139, 140, 147, 148, 152, 167, 173, 177, 183, 186, 192, 200, 201, 230
Crooked Creek settlement: 158
Crooked Creek Valley: 7, 17, 21, 101, 147, 171
Crooked L: 196, 197
Custer: 95

D

D Cross Outfit: 149, 179
Danills, Ben: 89
Day Outfit: 197
Deardoff: 116, 138
Decow, George: 21
Deep Hole: 118
Defreese, Billy: 22, 23
Delmar, John: 180, 181
Desmarien, E.P.: 76
Dickey Ranch: 164
Didger, Larry: 88
Dixon, Billy: 88
Dodge City: 7, 9, 12, 13, 17-19, 22, 28, 31, 35-38, 47, 48, 50, 51, 61, 62, 66, 84, 86-89, 91, 93-96, 99, 101-103, 107-109, 115-119, 121, 122, 125-127, 135, 138, 140, 141, 147, 148, 156, 158, 160, 163, 167, 171, 184, 185, 190, 223, 224, 240, 241
Dole, Frank: 181
Drapers: 88, 96

Driscoll: 103, 107
Drum, Henry: 218, 219
Dudley Ranch: 208

E

Earp, Wyatt: 88
Eddy, George: 88
Egg Half Diamond Ranch 116, 131, 138
Eliason: 19
Emerson, George: 21, 88, 96
Emerson Grove: 21, 28, 43
Englewood, KS: 74, 76, 88, 103, 168, 185, 186, 189, 218, 230, 231, 232
Ennis: 186
Erikson: 19
Evans: 186

F

Farris: 146
Fitzgerald, Herb & Charley: 116, 139
Five Mile Creek: 88
Fleetwood: 76
Fleming, Zack: 177
Fort Dodge: 17, 95, 125
Fort Elliott: 48, 117
Fort Hays: 95
Fort Lyon: 121
Fort Supply: 165, 207
Fosher: 186
Fowler, KS: 43, 226
Fox: 186
Frazier, John: 102
French, Captain: 109, 158
French: 19
Fuller, Rock: 213, 214

G

GG Ranch: 137, 139, 143, 172, 217, 220, 221, 222
Gay, E.L.: 187
Glenn, Johnny: 178
Goodnight, Charles: 7
Gorham Ranch: 155
Great Bend: 115

Greenhorn Mountain: 120, 239
Grimmer, Charley: 127
Grove: 186
Gruenken: 58

H

H Ranch: 191
Hackberry Creek: 126
Happy Jack: 159, 160, 166
Hardesty Brothers: 126, 197
Harlan, Lee: 97
Harrington, Al: 177
Harwood, W.I.: 139
Hash-knife: 223, 224, 227
Hay Meadow Massacre: 208
Hays: 138, 139
Haywood: 19
Healy, Frank: 76, 154
Healy Brothers: 117, 152, 172, 197, 209, 228
Healy Ranch: 154, 155, 172, 200
Healy Trail: 75, 96, 187, 230
Heart Jinglebob Brand: 217
Heinz, Charley: 117
Henry, Dutch: 88
Hensley: 43
Hobbs, Win: 45, 66
Hobbs, Fred: 192
Hog Creek: 186, 207, 208
hokey-pokey: 219
Holbrook: 223, 224
Holloway: 116, 131, 133, 134, 137-139
Holman, Don: 199
Holstein, Sim: 177, 219
Holtz, Pete: 199
Hoo-doo Brown Road Ranch: 148
Hoover, George: 88
Horse Creek: 116, 194

Horse Flat: 192
Howard, Lee: 90, 179
Huerfano: 120
Hugh: 116
Hull, Wes: 181

I
Innes, John: 67

J
J.B. Outfit: 113, 127
Jetmore, KS: 103
Johns Creek: 155, 174, 177, 182, 226
Johns, D.S.: 174
Johns, S.S.: 197
Johnson, Andy: 88, 89
Johnston, Thomas "Slick": 45, 68, 73, 81, 191, 192, 201, 202
Jones & Plummer Trail: 17, 27, 43, 48, 52, 84, 96, 117, 134, 148
Judy: 186
Judy, Fannie: 81

K
K Ranch: 147
K.K. Outfit.: 197
Keech, Archie: 97, 241
Kelley, "Dog":88
King, Joe: 117, 177, 179, 199
Kiowa Creek: 76, 102, 117, 155, 177, 215
Kollar, George: 102
Kollar, Hi: 102, 179
Kollar, Henry:102
Kramer and Sons: 126
Kramer Ranch: 148, 197
Kyger, Sam: 94, 102, 178
Kyger Creek. 102

L
LaForce, Perry: 183
Lake View: 67
Lamblin, Pete: 168
Lane, Pete: 117, 134

Langston, Jimmy: 88
Laurel Leaf Ranch: 210
Lemert, Labon: 52, 53
Lewis: 11
Liberal, KS: 177
Lockhart: 15, 19
Lovell: 34, 101, 102

M
M C Brand: 146
M Ranch: 117
Mackey, Dave: 53, 131, 175, 183
Major B: 192-195, 204
Maphet: 186
Maple: 186
Martz, M.C.: 51-53, 183
Masterson, Jim: 88
Mathers, Dave: 88
Maverick: 130
McCanly: 19
McCarthy, Dr.: 88
McCoy Outfit: 117
McGoven, "Irish": 191
Meade, KS:29, 57, 66, 69, 74, 77, 108, 117, 139, 167, 195, 209, 230
Medicine Lodge: 149
Meierdierks, Gerd: 222
Mexican Arroya: 135
Mexico Creek: 117
Meyers: 186
Miller, Nimrod : 120
Milligan: 19, 36
Miner, Lee: 59
Moore: 186
Morgan, Joe: 148
Mount Jesus Trail: 155
Mueller, John: 125
Mulberry Creek: 13, 95, 115, 137
mule skinner: 97
Murphy, Frank: 53
Musset, Vash: 177

N
Neutral City: 187, 207, 209

New Mexico: 6, 17, 68, 79, 91, 92, 94, 108, 122, 123, 141, 171, 183, 192, 217, 240
Nixon, Tom: 88
No Man's Land. 5, 6, 53, 68-70, 73, 185, 203, 208-210, 217
Noble, Charley: 177
Norman: 19

O
O'Louglin, John: 127
O'Neil, Hugh, Henry, Paddy and Jim: 116, 127
Odee: 183
Oil Can Brand: 146
Over, John: 117, 190
Over Ranch: 190

P
Palo Duro Creek: 96, 111, 126, 178
Parks: 186
Patterson: 150, 151
Peacock: 88
Pearlette : 19
Peckham, John: 68
Pemberton: 186
Penrose, Harvey: 14, 19, 20, 23
Penrose, Isaac: 20
Penrose, Will: 20
Perry, R.K.: 53, 116, 136, 195, 197
Perry, George: 183
Peterson: 19
Petty: 186
Pietz: 186
Pike, Zebulon: 92
Point of Rocks:113, 127, 130, 151, 239
Ponca County: 149
Powers, Fred: 183
Prairie Dog Dave: 84
Protection, KS: 102, 155
Pueblo, Colorado: 119, 128, 149

R

Raton Pass: 92, 93
Red River: 99, 149
Reep: 19
Remuda Creek: 39, 49
Rexford, "Spike": 21
Reynolds, P.G.: 36, 88, 117, 118
Rhodes, N.J.: 53, 116, 211
Rhodes, Wiley: 225
Rio Grande R.R.: 120
Roberts, Jack: 77
Roberts, Gus: 181, 182
Roberts Family: 186
Roberts, Wm.: 217
Robin Hood: 166
Rockfeller: 191, 192
RS Outfit: 117, 166, 210
Ruble, Smith: 152, 177, 183

S

San Francisco Creek: 126
Sand, John: 178
Sand Creek: 22, 103, 108, 177
Sand Wells: 127, 150, 151
Santa Fe, NM: 91-94
Santa Fe Railroad: 11, 75, 84, 94, 95, 185
Santa Fe Trail: 91, 93, 94, 229
Sawlog: 125
Schultz: 23
Scully, Martin: 146, 147
Sewall, Bell: 52
Shaefer: 186
S-Half-Circle: 126
Sharp's Creek: 117, 150, 155, 177
Short, Luke: 88
Siebenthaler, Wil: 44
Slover, Jessie: 68
Smoky River: 115, 127, 128, 145, 203
Snake Creek: 185
Sourbier: 19

Spearville, KS: 84
Spencer, Frank: 127, 152
Spurgeon: 186, 237
Stafford, Frank: 97
Steele, F.M.: 81, 202
Steele, N.E: 139
Stevens County: 208
Stone Calf: 164, 165
Stone School: 68, 80, 81
Streter, Walter: 88
Supply Trail: 96, 118

T

T6 Outfit: 195
T Pete: 181
Taft: 186
Taintor, Fred: 75, 116, 139, 140, 142, 144, 146, 167, 168, 177, 184, 186, 189, 190, 192, 199, 202, 206, 210, 217, 218, 223, 224, 227
Taintor Creek: 116, 139, 144, 168, 209
Tascosa Trail: 96
Three Seven Ranch: 126
Through Trail: 95, 96, 101, 103, 116-118, 121
Trinidad, CO: 122
Tuttle Trail: 96, 163
Tyson, O.N.: 128, 164

U

Updegraff, Alf: 88

W

Wagon Bed Springs: 127, 150, 152, 182
Webster: 88
Whitaker: 186
Williams, Sam: 33-36, 108
Windor, Labe: 147
Winer, Leo: 97
Wolf Creek: 95, 109, 115, 130, 147, 152, 154, 168, 177, 179, 190, 213

Woodward, OK: 167
Worth: 19
Wrangler: 128
Wright, Bill: 105
Wright, Bob: 88, 224
Wright: Howard: 147
Wurdeman: 57

X

XI Ranch: 183, 193, 197, 217

Y

York: 88, 96, 97

Z

Zula, TX: 127